Peter Rudolph

Crystal Growth Fundamentals

Also of interest

Crystal Growth Fundamentals
Peter Rudolph

Kinetics of Crystallization, planned for 2026
ISBN 978-3-11-171416-5, e-ISBN 978-3-11-171422-6
Transport of Heat and Mass, planned for 2027
ISBN 978-3-11-171417-2, e-ISBN 978-3-11-171429-5

Also available as a SET:
ISBN 978-3-11-172729-5

Physical Metallurgy.
Metals, Alloys, Phase Transformations
Vadim M. Schastlivtsev, Vitaly I. Zel'dovich, 2022
ISBN 978-3-11-075801-6, e-ISBN 978-3-11-075802-3

Thermodynamics.
From Energy and Entropy to Heat Transfer and Phase Transitions
Rainer Müller, 2025
ISBN 978-3-11-031018-4, e-ISBN 978-3-11-031020-7

Thermal Analysis and Thermodynamics.
In Materials Science
Detlef Klimm, 2022
ISBN 978-3-11-074377-7, e-ISBN 978-3-11-074378-4

Peter Rudolph

Crystal Growth Fundamentals

Part I: Thermodynamics of Crystallization

DE GRUYTER

Author
Prof. Dr. habil Peter Rudolph
Crystal Technology Consulting
http://crystal-technology-consulting.de/
Helga-Hahnemann-Str. 57
D - 12529 Schönefeld
rudolph@ctc-berlin.de

ISBN 978-3-11-171105-8
e-ISBN (PDF) 978-3-11-171116-4
e-ISBN (EPUB) 978-3-11-171118-8

Library of Congress Control Number: 2025932387

Bibliographic information published by the Deutsche Nationalbibliothek
The Deutsche Nationalbibliothek lists this publication in the Deutsche Nationalbibliografie; detailed bibliographic data are available on the internet at http://dnb.dnb.de.

Cover image: 5-inch GaAs vapor pressure controlled Czochralski (VCz) crystals of near-stoichiometric composition grown in a thermodynamically pre-determined arsenic atmosphere by the author's former team at the IKZ Berlin (*photo from the author's personal collection*)
Typesetting: Integra Software Services Pvt. Ltd.

www.degruyter.com
Questions about General Product Safety Regulation:
productsafety@degruyterbrill.com

To my wife, Petra

About the author

Peter Rudolph is a professor emeritus of crystallography and crystal growth. He was born in Germany in 1945. He studied and defended his Ph.D. at the Technical University of Lviv (Ukraine). After his long-year stay at the Humboldt University of Berlin and visiting professorship at the Tohoku University in Sendai (Japan) he was employed at Leibniz Institute for Crystal Growth in Berlin. Since 2011 he is active as consultant for crystal growth technology. He acted as lecturer of crystal growth fundamentals in nearly 30 countries. He was a member of the Executive Committee of International Organisation of Crystal Growth (2004–2011) and the president of the German Association of Crystal Growth (2010–2011). He was awarded the Medal of the German Association of Crystallography in 1990 and Innovation Prize Berlin-Brandenburg in 2001 and 2008. He is one of the editors of the three-volume *Handbook of Crystal Growth* (Elsevier 2015) and seven other books on crystallization. His professional literature includes over 300 papers and book contributions as well as 35 patent publications. In 2023 he was awarded the Laudise Prize of the International Organization for Crystal Growth for exceptional achievements in the field of crystal growth, particularly with regard to technological solutions.

https://doi.org/10.1515/9783111711164-202

Contents

About the author —— VII

Prologue —— XI

Part I: **Thermodynamics of crystallization**

1 Introduction —— 3
1.1 The importance of thermodynamics for crystal growth —— **3**
1.2 Equilibrium and nonequilibrium thermodynamics —— **4**

2 The potential of Gibbs —— 7
2.1 The principle of Gibbs free energy minimization —— **7**
2.2 Phase transition and enthalpy of transformation —— **8**

3 Phase equilibrium and phase diagrams —— 16
3.1 Two phases of one component —— **16**
3.2 Two phases of two (or more) components —— **19**
3.2.1 Basic principles —— **19**
3.2.2 Ideal mixed systems —— **26**
3.2.3 Real mixed systems —— **31**
3.2.4 Selected binary phase diagrams —— **40**
3.2.4.1 Mixed crystals with nearly ideal solid solution —— **40**
3.2.4.2 Systems with congruent melting compounds —— **42**
3.2.4.2.1 Stoichiometry and region of homogeneity —— **42**
3.2.4.2.2 The problem of incongruent evaporation —— **48**
3.2.4.2.3 The control of stoichiometry via in situ vapor–liquid–solid-phase equilibration —— **53**
3.2.4.2.4 Vapour pressure controlled Czochralski growth without liquid encapsulant —— **63**
3.2.4.3 Systems with incongruent melting compounds —— **65**
3.2.4.4 Systems with solid–solid transition —— **67**
3.2.5 Ternary systems —— **70**
3.2.6 The thermodynamic equilibrium segregation coefficient —— **73**
3.2.6.1 Segregation at melt–solid phase transitions —— **73**
3.2.6.2 Segregation at solution–solid phase transition —— **84**
3.2.6.3 Segregation at vapor–solid phase transition —— **88**
3.2.6.4 Segregation at vapor–liquid-phase transition —— **89**

4 Surfaces, phase boundaries, and interfacial effects — 91
4.1 Determination of the surface free energy — **91**
4.2 Gibbs–Thomson equation — **95**
4.3 Equilibrium shape of crystals — **98**
4.4 Selected effects of surface energy on crystal growth processes — **106**
4.4.1 Growth angle at crystal pulling from the melt — **106**
4.4.2 Faceting and ridge formation at melt growth — **108**
4.4.3 Surface energetic effects at epitaxy — **112**
4.4.3.1 Surface reconstruction — **113**
4.4.3.2 Ordering effects in mixed semiconductor thin films — **117**
4.4.3.3 Surface patterning by self-assembling — **120**

5 Deviation from equilibrium — 125
5.1 Driving force of crystallization — **125**
5.2 Nucleation — **130**
5.2.1 Homogeneous nucleation — **133**
5.2.1.1 Classical approach — **134**
5.2.1.2 Nonclassical concept — **143**
5.2.2 Heterogeneous nucleation — **145**
5.2.2.1 Basic considerations — **145**
5.2.2.2 Application in epitaxial processes — **149**
5.2.2.3 Nonequilibrium nucleus distribution — **152**
5.2.3 Uncontrolled nucleation in crystal growth containers — **154**
5.2.4 Precipitation in cooling crystals — **156**
5.3 Ostwald ripening and grain coarsening — **158**
5.4 Nonequilibrium (kinetic) phase diagrams — **161**
5.5 Nonequilibrium thermodynamics: basic principles for crystal growth — **163**

6 Conclusions — 173

Recommended literature — 175

Abbreviations — 179

Spec boxes — 181

Index — 183

Prologue

Crystal growth is a fascinating field that is more and more needed in almost all walks of life. Unfortunately, this fact is hardly known. Why? On the one hand, this task is typically associated with the production of artificial gemstones only (often, when I introduced myself to previously unknown persons, I was asked whether I am able to produce a diamond or even a “crystal vase”). Indeed, meanwhile relatively small but perfect diamond crystals can be grown. However, they are nowhere near as interesting for the gemstone trade as they are for high-power microelectronics. Substrates made of single crystalline diamond show highest thermal conductivity, enabling an enormous dissipation of process heat. And so we have already reached one of the most important application fields of artificial crystals (the “crystal vase” made from glass and, thus, nothing to do with crystal growth we quickly put aside). Nanocrystals, high-quality epitaxial thin films, and bulk crystals are of high importance for micro- and optoelectronics, photonics, computing, communications, energy saving and storage, radiation generation and detection, medicine, biotechnology, homeland security, and so on.

On the other hand, crystals are usually not recognizable on the exterior of a technical equipment or device. Mostly, they are small-sized centerpieces of a device or the basic slice of a circuit or in a process machine entirely covered by protective casing and conductors. For instance, today each automobile, computer, cell phone, CT scanner, or tool for laser operations is equipped with devices made of various crystalline pieces. A large charge from them is directly visible as a light-emitting pixel display. Rarely are bigger make of a crystal in the form of quite large silicon wafers to be seen, as in solar cells. Or, who knows right away that the huge lenses with a diameter of 300 mm in lithography systems of ultra-short wavelengths (so-called waver steppers) are made from CaF_2 single crystals?

However, the most problematic aspect of the general lack of knowledge about single crystals and crystal growth proves to be the absence of education and inadequate media presentation. Usually, artificial monocrystals are treated as of secondary importance within the framework of physics, chemistry, and materials science. Mostly, their crystallographic and growth principles as well as broad applications are outlined in introduction only. Obviously, this has to do with the fact that the mastery of crystallization and epitaxial processes on highest level as possible requires a profound interdisciplinary knowledge that combines physics, chemistry, mathematics, crystallography, materials science, electronics, automation, engineering, and so on. Nowadays elementary and special knowledge of biology and medicine also belong to it. These facts upgrade the wide field of crystal growth to a challenging quasi-self-contained interdisciplinary branch of science.

Unfortunately, the current level of related academic education does not meet this challenge. Even during the last decades, training in this field is decreasing. Looking back on the international situation at the turn of the century, compared with today, many more academic departments and research laboratories dealt with the fundamentals, ex-

https://doi.org/10.1515/9783111711164-204

periments, and technology developments of crystal growth. At present, there are only two autarkic institutes for crystal growth in Berlin (Germany) and Kharkiv (Ukraine). Of course, additionally, numerous excellent institutes of materials science with partial orientation on crystal exist, as in the USA, Japan, China, Switzerland, South Korea, Singapore, and India. However, where and how the young academics having a special knowledge for their needed crystal growth research are educated? I would like to remind the readers that until the German reunification in two self-dependent departments of crystallography at the universities of Berlin and Leipzig were trained "Diploma Crystallographers" with comprehensive knowledge on crystal growth and analysis. In addition to the basic courses of physics, chemistry, and crystallography, the students attended profound lectures on thermodynamics and kinetics of phase transition, crystal growth fundamentals and technologies, defects, and crystal applications. I by myself lectured for many decades in Berlin such disciplines and supervised numerous PhD students on crystal growth. Unfortunately, such goal-oriented education no longer exists. After one of my popular scientific lectures on crystal growth and application in 2011 in a high school, one of the enthusiastic pupils asked me "where can I study this fantastic subject?". Sadly, my answer was "nowhere. . .". Today, as anywhere in the world, a young scientist who is assigned a task on growth and analysis of a new crystal material, nanocrystal or epitaxial layer must familiarize themselves in a time-consuming independent study via textbooks and internet data, however, without any seminar-style discussions and practical trainings. Fortunately, there are some occasionally organized international and national schools on crystal growth providing over a period of about 1-week fundamental lectures. Over the years, the high interest and visitor volume at such training courses are the evidence that the demand of specialists contradicts the actual situation of missing academic education yet. Therefore, it is also clear why according to the statistics and scores of publication databases even reviews and editions on crystal growth and defect formations show exceptionally high read rates.

How did it come to such contradictory situations? First, there is a widespread tendency that the fundamentals and technological means of crystallization phenomena and their control are more or less already solved. Of course, over more than a half century of the development of crystal growth technology, most of the important mechanisms have become well understood. But that is not to say that all new challenges and arising problems are already mastered. For instance, although we understand fully the conditions under which morphological growth instabilities occur, it is still not possible to obtain the detailed parameters that permit the production of large, homogeneous alloy (mixed) single crystals consisting of two more components that would be invaluable as tailored substrates. For its future mastery, the crystallization process must be combined with newly developed automation programs. Further, twinning remains a serious limiter of yield in the growth of single crystals with low stacking fault energy. It seems to be due to the appearance of facets but we do not exactly understand the decisional origin yet. Then, the optimum growth conditions of high-quality large-sized GaN crystals, being extraordinary important for optoelectronics and future high-power devices, are not yet mastered. Also, the production of reproducible CdTe single crystals as radiation detectors for medical diagnostics by computer tomography needs still the minimization of diverse growing-in defect phenomena. The many other examples include growth of functional materials for electromechanical energy harvesting, monocrystals for high-efficiency hydrogen storage, periodically structured crystals for photonics, and perfect monocrystalline lenses for laser-excited fusion energy. Many further arguments can be extended to the branches of nanocrystals and epitaxial thin films too.

Another reason of the current public drop in activity levels in crystal growth study and education is due to the high-tech and strategic character of single crystals and advanced thin-film configurations. Meanwhile, there are only few remaining highly developed industrial producers with market leaderships which are increasing isolate itself, and thus excluded from the public sphere with the aim to ensure the dominating role of international competition. Actually, the number of speakers from industry at conferences of crystal growth is significantly reducing. As a result, numerous development problems, especially of technological character, do not make their way to the outside but are developing further in own R&D laboratories without access for public academic researchers. In part, this is understandable. However, it does not solve the question of where these companies draw their highly qualified new employees? Is it really sufficient to transmit the working knowledge behind the closed doors only? Taking, for example, the scientific penetration of current related patent publications (now the almost only allowed communication type of the industry), the precision of which leaves often much to be desired.

Finally, until now the crystal growth community has hardly any financial and political supporters. Despite the long-standing efforts of the International Organization of Crystal Growth (IOCG) and numerous related national associations, the general attention is decreasing among the governments. In recent years, numerous institutes and laboratories specified on crystal growth and even some national crystal growth communities have disbanded. The common opinion is that the mission and necessity of crystal growth can be settled by the firmly established areas of physics, chemistry, and materials science. However, until now this succeeded only to a limited extent.

Whatever the development, it is my deep intention to contribute to maintaining and reinforcing the knowledge of fundamentals of crystal growth even to the worldwide young researchers. Therefore, I publish now my related lectures that I hold over 20 years at the Humboldt University in Berlin and subsequently in many universities, institutes, and companies in about 30 countries. Moreover, I was teacher of 7 international summer schools of the IOCG and about 20 crystal growth courses of the International Union on Crystallography (IUCr) and diverse national communities. Now is the time to commit my lecture texts and slides to publish. Of course, during all of this time of my knowledge transfer I was always tried to keep abreast with the latest state of development and to bring the present book to the current state of the art.

I would like to emphasize that there is already a wide collection of excellent reviews and textbooks on crystal growth fundamentals. Most of them I studied as base material for my lectures and refer to them at the end of my lecture parts. I recommend emphatically using these papers and books for in-depth knowledge. However, in my opinion, a coherent textbook series of introductory character on fundamentals of crystal growth and defect control even for newcomers and further training is still missing. Of course, the outstanding three volumes are Elsevier's *Handbook of Crystal Growth* of first edition (1994) and second edition (2015), and Springer's *Handbook of Crystal Growth* (2010). Though the numerous chapters of these editions have been written by various authors of different scientific levels requiring in many cases high expertise, numerous redundancies are unavoidable. In comparison, it is intended that the overall image of the present lecture collection is coherent and particularly suitable for beginners. No detailed presentations of higher mathematics are provided (some special important derivations are given in gray backing Spec boxes). As a special feature, figures are prepared in the graphical form identically with my lecture slides often combining sketches and images with the general formulas. They also contain the corresponding authors and literature references for further studies.

The starting slide is shown in Fig. P.1. It shows a sketched *crystallization front*, also named *solid–fluid interface*, propagating with normal *growth velocity* v. thermodynamically, v depends on the *driving force of crystallization* $\Delta\mu$ being the potential difference between the phases, here proportional to the difference between the equilibrium temperature (T_{eq}) and undercooled value (T_{IF}) at the interface. However, be-

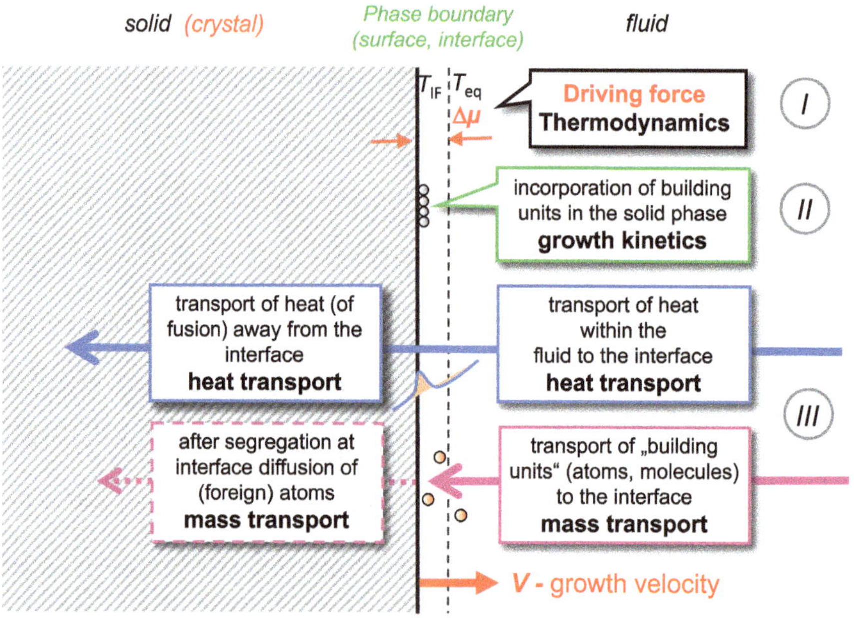

Fig. P.1: Partial processes determining crystal growth: thermodynamic (I), kinetic (II), and transport of heat and mass (III).

cause the thermodynamics is not able to impart the crystallization processes at the growing interface in microscopic details, the branch of *kinetics* becomes involved. It shows the various interface nature from atomistic view and its growth mode as a function of it atomically smoothness and roughness as well as of the presence of defects and foreign atoms. Finally, each crystallization requires temperature and concentration (pressure) *gradients*. This is due to the necessary control of the transport of heat, especially the generated heat of fusion away from the interface. Additionally, the transport of crystal *building units* (atoms, molecules, and dopants) toward the growing interface is required. At the same time, undesired foreign atoms (impurities) should be repulsed at the growing interface as effectively as possible. As can be seen, the crystal growth processes prove to be varied and versatile, which requires a comprehensive study. It can be stated that the *fundamentals of crystal growth* are based on three factors:
(i) *thermodynamics* (of phase transition),
(ii) *kinetics* (of crystallization processes), and
(iii) *transport* (of heat and mass).
This is summarized in Fig. P.2.

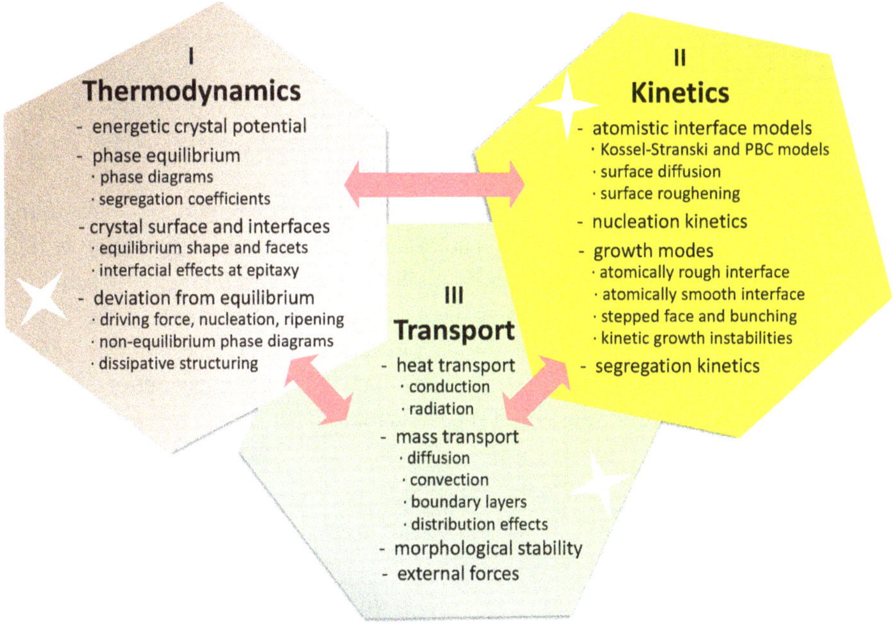

Fig. P.2: Three pillars of fundamentals of crystal growth.

Some decisional topics of each field to be treated by the following lectures are added. Of course, all three basics are closely interrelated. Nevertheless, their individual treatment has proven to be the best and most logical for teaching effectiveness. In a fourth lecture part, the origins of

(iv) *crystal defect* formations and their mastery are planned to add. Their exact understanding is only possible on the basis of the three crystal growth fundamentals (i)–(iii).

My lecture parts correspond to this chronology.

I am very grateful to all previous national and international students for their participation and interest in discussions in my lectures. I also have to thank all my former research co-workers and teaching colleagues at the universities and institutes where I was employed as well as numerous members of the German Association of Crystal Growth and IOCG. Their widespread support, critical comments, and recommendations have made an important contribution to the continuous improvement of my lecture level. I would particularly like to mention two scientists who shaped my own professional training in the field of crystal growth fundamentals essentially. On the one hand, this is Dr. Lars Ickert, who introduced me by his excellent lectures into the fundamentals of phase transition, nucleation, and epitaxy as I started to work at the Department of Crystallography, of the Humboldt University, as young postdoc. On the other hand, this is Prof. Alexander A. Chernov who fascinated me with his fabulous training course on crystal growth fundamentals during the former international summer school on crystal growth in Varna (Bulgaria). Also his comprehensive chapter on crystal growth in Springer's *Modern Crystallography III* (1984) provided me the first overall overview of this enthusiastic scientific field.

It is a great pleasure for me to thank the Walter de Gruyter GmbH for enabling the publishing of my long-lived lecture courses on fundamentals of crystal growth in book form and, particularly, for including my lecture slides as figures in original design. My special thanks go to the Senior Acquisitions Editor Physical Sciences Kristin Berber-Nerlinger, the Editor Ute Skambraks, and the Senior Project Manager Kowsalya Perumal for the excellent cooperation, consultancy and their great effort with the print preparation of text, formulas and reproductions.

Peter Rudolph
Schönefeld, May 2025

Part I: **Thermodynamics of crystallization**

1 Introduction

1.1 The importance of thermodynamics for crystal growth

As much as we shy away from thermodynamics, we cannot do without its basic importance for crystal growth and, besides, we will see below that we can even enjoy it.

To crystallize an element, compound, mixed, or heterogeneous material artificially, more exactly, to arrange a given kind or combination of atoms or molecules in a well-ordered crystalline structure, one must first start with the transfer of the raw material (feedstock) into a still unordered (sometimes partially preordered) fluid phase (nutrient) without fixed particle correlation, such as melt, gas, or solution, in order to force the atoms or molecules into a strongly structured assembly by applying a sophisticated operation of solidification. Such a process requires the exact choice of the starting material and professional navigation of temperature, pressure, and quantity of involved chemical components. With this, we are already in the middle of thermodynamics, having to do with thermodynamic parameters such as T – temperature, p – pressure, and n_i – component quantity, as well as system variables such as V – volume, E – energy, H – enthalpy, and S – entropy, for example, which one has to control by certain crystal growth conditions, like supercooling, supersaturation, gradients, cooling rates, and time programming.

Consequently, thermodynamics is an important practical tool to understand crystal growth. It helps to: (i) understand the material properties, its phase relations, existence region, and stability behavior under growth conditions; (ii) find out the most effective phase transition and select the related optimum growth conditions; (iii) determine the driving force of crystallization; (iv) estimate the nucleation mode, crystallization velocity, and expected crystal morphology; (v) appreciate the distribution of dopants and impurities along the growing crystal; and (vi) establish the best measures of process control. In short, no technological optimum can be found without considering thermodynamic relationships. Thus, thermodynamics belongs not only to the basic theoretical knowledge of each crystal growth but also challenges its practical dexterity and clever modern mastering of process control and automation.

However, we must always bear in mind that thermodynamics is a macroscopic science and, therefore, of *phenomenological character* only. It deals with average changes taking place among large numbers of atoms or molecules. It shows solely macroscopic start and end states, phase relations, tendencies, and directions but not the pathway in detail, as well as the microscopic steps of atomic size during the building of a crystalline structure. In short, thermodynamics gives answers to the questions "why" and "when", but it cannot explain the "how". Responsible for the latter is the field of kinetics – the inseparable partner within the spectrum of the "science of crystal growth" to be treated in my lecture part II.

https://doi.org/10.1515/9783111711164-001

1.2 Equilibrium and nonequilibrium thermodynamics

Let us assume an idealized "closed physical system" without any energy exchange with the environment, consisting of two heating zones arranged one after the other, with thermal isolation in between, quite reminiscent of a crystal growth furnace. Only zone 1 is hot at temperature T_1, while zone 2 is cold at T_2, as sketched in Fig. 1.1a. Such a situation would represent an orderly system because of the clear distinctiveness of both regions. After removing the thermal isolation between both zones, according to the *second law of thermodynamics*, a heat exchange dQ proceeds from the hot zone 1 toward the colder zone 2 by the inner heat flux $j_{T\ int}$, producing entropy dS according to the Clausius theorem

$$dS \geq dQ \left(\frac{1}{T_2} - \frac{1}{T_1} \right) \geq 0 \tag{1.1}$$

until the thermal balance $T_1 = T_2 = T$ and $dQ/T = 0$ occurs. In the sense of *equilibrium thermodynamics*, this final situation means that the system entropy $S = Q/T$ has reached its maximum, or the entropy production has become zero because the former orderly system has passed into a disorderly state of thermal equalization as follows:

$$dS = dQ/T = 0. \tag{1.2}$$

a closed system b open system

Clausius theorem:

$$dS \geq dQ \left(\frac{1}{T_1} - \frac{1}{T_2} \right) \geq 0$$

$$dS = dS_{\text{ext}} + dS_{\text{int}} > 0$$

$$S < S_{\text{max}} \quad \text{contiuous entropy production !}$$

$S \rightarrow \text{max}, dS = dQ/T = 0$
highest degree of disorder

$P_s = dS/dt > 0$ *e.g. when*
ordering **possible** $dS_{\text{ext}} > dS_{\text{int}}$

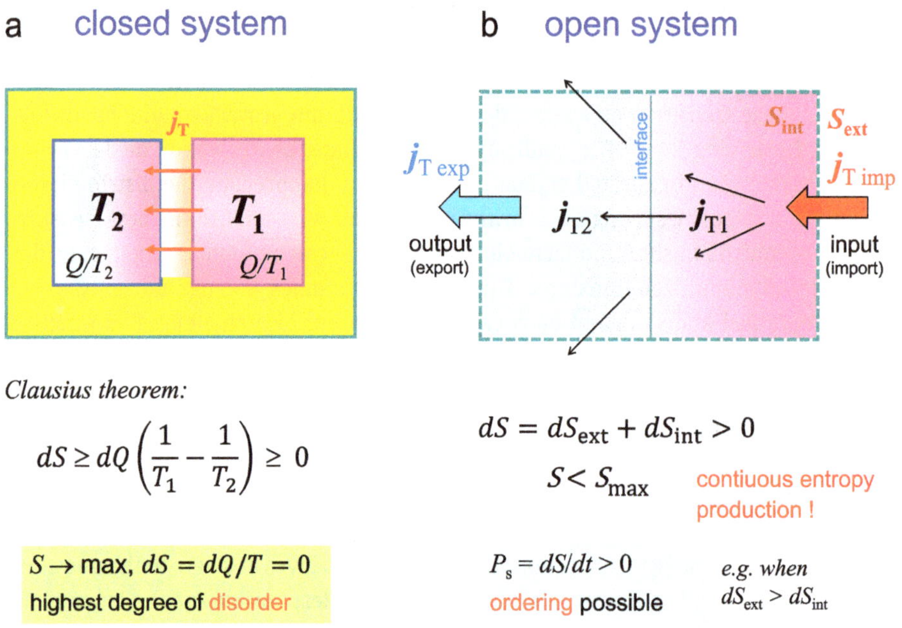

Fig. 1.1: Closed (a) and open system (b) described by equilibrium and nonequilibrium thermodynamics, respectively.

From now on, any small deviation (fluctuation) from the equilibrium will always be "reversibly" balanced.

In comparison, an "open physical system" is characterized by a continuous exchange of energy with the environment (Fig. 1.1b). Let's look again at the previous example of a two-heater furnace. In order to maintain a steady temperature difference between zones 1 and 2, almost to establish a temperature gradient, one has to compensate for the real constant loss (export) of heat through the chamber wall by a steady supply (import) of heat energy into the hot zone. Then the internal thermal flow $j_{T\,\mathrm{int}} = j_{T1} + j_{T2}$ from zone 1 toward zone 2 is extended by the import and export amounts $j_{T\,\mathrm{imp}}$ and $j_{T\,\mathrm{ext}}$, respectively. In such a case, the entropy production, consisting now of internal dS_{int} and external dS_{ext} terms, is never completed and always remains greater than zero:

$$dS = dS_{\mathrm{ext}} + dS_{\mathrm{int}} > 0 \tag{1.3}$$

What this means is that in a "thermodynamically open system," the entropy never reaches the maximum value and is always

$$S < S_{\max}. \tag{1.4}$$

Due to the unattainability of thermal balance, such a system like in Fig.1.1b is in accordance with *nonequilibrium thermodynamics*.

Strictly speaking, crystal growth and phase transition also occur under nonequilibrium conditions and belong to an open (irreversible) thermal system where steady internal and external transports of heat take place. Hence, the "building blocks" of a growing crystal (atoms and molecules) follow the path from fluid to solid phase in the sense of an irreversible flow.

Nevertheless, crystallization is generally treated in the sense of thermodynamic equilibrium. Why? Firstly, most crystal growth processes require a relatively small deviation from equilibrium, characterized by an almost perfect balance between the incorporation and removal of building blocks at the interface between solid and fluid phases. Secondly, the time for the macroscopic processes of net transfer of heat and mass between the nutrient and solidifying phase is much longer than for the microscopic events in each phase, such as atomic fluctuation rates and transition kinetics among the atomic states. As a result, the atomic fluctuations in the system determine the macroscopic system parameters via a quasi-statistical average. Hence, the equilibrium treatment of crystal growth proves to be a sufficiently good approximation that will be applied at the most following treatises too.

However, it is necessary to note that the current theoretical treatments of crystal growth are increasingly using the principles of "linear nonequilibrium thermodynamics" to describe more exactly the processes of crystallization. Especially, the complex defect dynamics in growing crystals is partially influenced by the phenomenon of irreversible thermodynamics. As was already mentioned above, a basic characteristic of irreversible processes is the continuous production of entropy $P_S = dS/dt$ during a given time t. The most "exciting" result of continuous entropy production is the system's ability to generate ordered *dissipative structures*. Even multiparticle systems, such as material phases, possess enormous structural reserves for self-organization. The total system entropy S_Σ now consists of several subentropic parts. In the case of crystal growth, these are the terms related to heat flow S_T, mass transfer S_i, friction S_η, and any internal (chemical) reactions S_Q:

$$S_\Sigma = S_T + S_i + S_\eta + S_Q + \cdots < S_{\max}. \tag{1.5}$$

According to the *evolution criterion* of *Glansdorff and Prigogine* in such an irreversible open system, any of the internal entropy parts can be removed. As a result, the sum of total entropy S_Σ decreases, and a higher ordered state will appear. For instance, well-known dissipative structures in crystallization processes are convective flow patterns in the fluid phases, strong-periodically cellular melt–solid interfaces, and dislocation cell patterns (some more details are given in Section 5.5).

But now, let us start with the basic phenomena of quasi-equilibrium thermodynamics. Note, we will not derive all relations, laws, and equations in detail. For this purpose, an enormous number of specialized textbooks, review articles, and treatises are available. In addition, it can certainly be assumed that every reader has a sufficiently good basic knowledge of general thermodynamics. Here, we will focus on the most relevant correlations for crystal growth only.

2 The potential of Gibbs

2.1 The principle of Gibbs free energy minimization

Deriving from the *first law of thermodynamics*, each system or state of matter (in crystal growth, the fluid and solid phases) possesses a conserved energetic potential G, which is available (free) at an excited condition under a given temperature T and pressure p, defined as Gibbs free energy (potential):

$$G(T, p) = U + pV - TS \tag{2.1}$$

which is the same as

$$G(T, p) = H - TS \tag{2.2}$$

where U is the internal energy, V the volume, S the entropy, and H the enthalpy.

From eq. (2.2), it becomes apparent that the convertible Gibbs free energy of a system consists of the internal part of enthalpy (first term on the right side), which is reduced by the bounded part (second term on the right side) representing the product of absolute temperature and entropy.

According to the general principle, *all states strive to minimize their free energy* by returning from their excited condition to a lower energy state (Fig. 2.1). Applied to the crystallization process, this means that the single-crystalline state is a normal one because the thermodynamic potential G of a solid phase becomes minimum if its building blocks (atoms, molecules) are perfectly packed in a three-dimensionally ordered crystal structure with regularly saturated atomic bonds. In sum, all atomic bonds together yield de facto the potential part H of the internal crystal energy U as follows:

$$U = H - pV \tag{2.3}$$

Therefore, the *process of ordering* of the atoms or molecules "in rank and file" over the course of regular crystallization minimizes the system's enthalpy:

$$H \rightarrow \min \tag{2.4}$$

and consequently, according to eq. (2.2), also the Gibbs free potential G of the crystal.

However, as was already discussed in Section 1.2, the entropy S strives to increase by inducing disorder. In other words, an ideally ordered crystalline state would imply a too limited entropy S. Thus, the minimization of Gibbs free energy is also proportionally realized by an oppositely directed force of increasing entropy, causing *certain disorder* according to the tendency

$$S \rightarrow \max \tag{2.5}$$

As a result, both effects of the opposite drive decrease the free energy, as it is expressed mathematically by the minus sign in eq. (2.2). That's the point of the basic

https://doi.org/10.1515/9783111711164-002

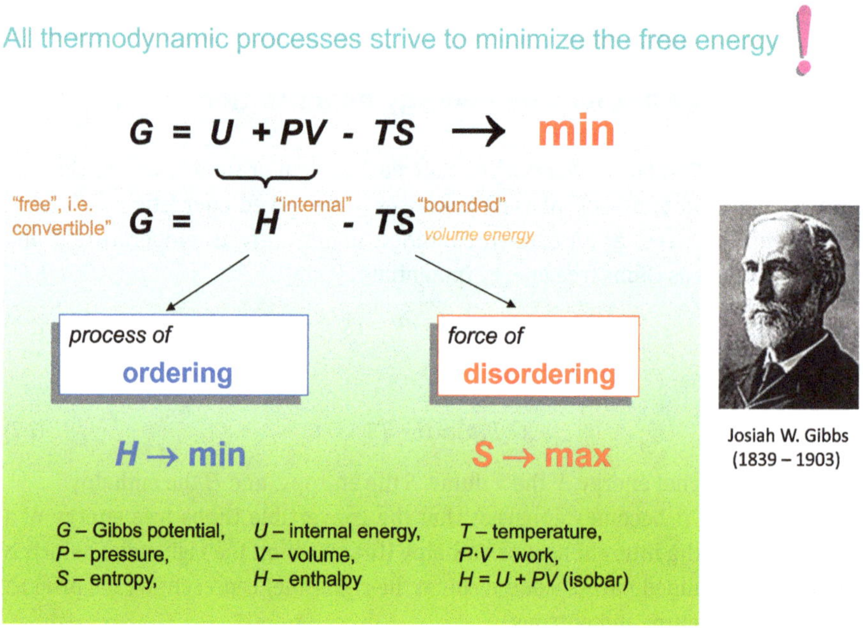

Fig. 2.1: The basic phenomenon of equilibrium thermodynamics (*the portrait of J.W. Gibbs is in the public domain*).

equation of Gibbs (Fig. 2.1). We will often meet its *dialectics* during our study of the phenomena of crystallization and defect generation. For instance, at thermodynamic equilibrium, there is no perfect crystal, but the crystallographic structure is somewhat affected by the misplacement of some atoms in a disordered manner referred to as *point defects* (see lecture part IV).

Note that the Gibbs free energy is a state function. Therefore, the change in Gibbs free energy, represented by ΔG, depends only on the initial and final states of the system and does not depend on the path by which the change has been carried out.

2.2 Phase transition and enthalpy of transformation

Each process of crystallization implies a phase transition. The building blocks are transferred from an unordered fluid phase (gas, melt, and solution) into an ordered one (crystallized solid). Additionally, there are also phase transitions in crystalline solids if they show a range of different crystal structures depending on temperature and pressure. Both crystallization and solid-state phase transfers are *first-order phase transitions* (Fig. 2.2). This means there is the coexistence of two distinct uniform phases that are stable at equilibrium and separated by a phase boundary, i.e., an *interface*. Close to equilibrium, the phases can still exist, one as thermodynamically sta-

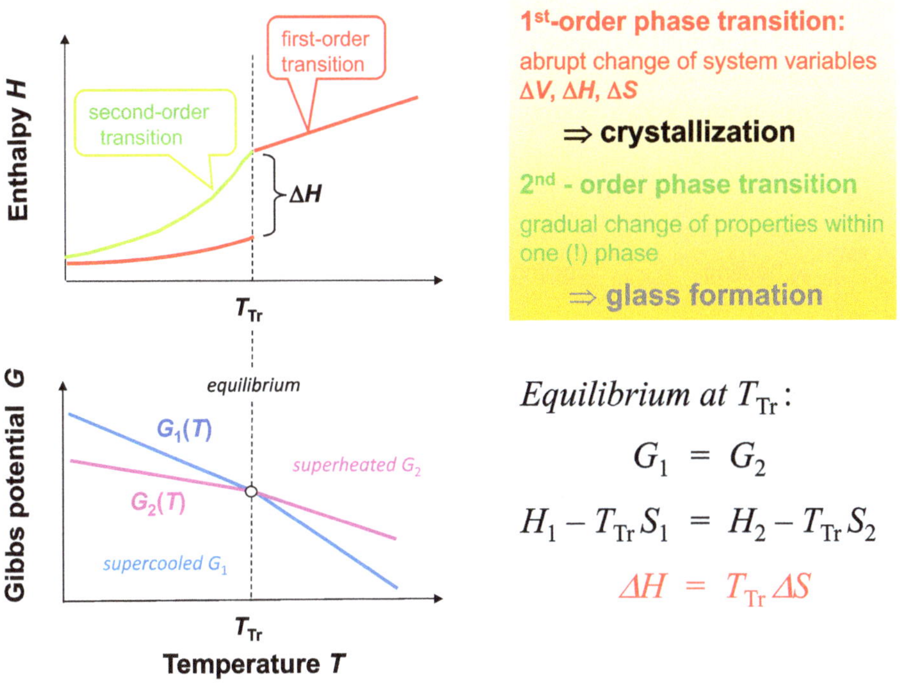

Fig. 2.2: First- and second-order phase transitions.

ble and the other as thermodynamically metastable. Depending on the direction in which the system is deviated from equilibrium, the metastable phase is superheated (undersaturated) or supercooled (supersaturated) with respect to the stable phase. At equilibrium of two phases 1 and 2, the first-order phase transition is characterized by an abrupt change of the thermodynamic variables as follows:

$$U_1 - U_2 = \Delta U,\ H_1 - H_2 = \Delta H,\ S_1 - S_2 = \Delta S,\ V_1 - V_2 = \Delta V \cdots \tag{2.6}$$

Thus, the amount of *heat (enthalpy) of transformation* ΔH, at constant pressure, is to be equated with the *latent heat of crystallization* $\mathcal{L}$, which is released or absorbed. Then, at the transition temperature, T_{Tr}, it becomes

$$G_1 = G_2 \tag{2.7}$$

$$H_1 - T_{Tr} S_1 = H_2 - T_{Tr} S_2 \tag{2.8}$$

$$\mathcal{L} \equiv \Delta H = T_{Tr}\,\Delta S \tag{2.9}$$

As it is well known, the specified value of $\mathcal{L} = Q/m$ (Q – heat required to completely effect a phase change, m – the mass of the substance) can be determined by differential thermal analysis and differential scanning calorimetry (DSC). Figure 2.3 shows the heating and cooling curves of CdTe and the thermal peaks of solid–solid phase transitions in CsCl. From the DSC curves of heat flux versus temperature or time, the transition enthalpy can be calculated by integrating the endothermal and exothermal peaks. Such measurements are essential for the determination of phase diagrams.

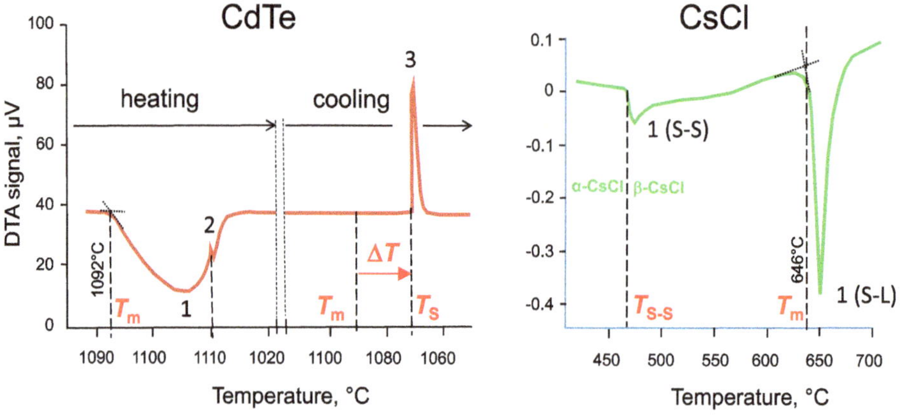

M. Laasch, P. Rudolph, own experimental data (1992)

Courtesy of D. Klimm, IKZ Berlin (2020)

DTA analysis of melting, crystallization and solid-solid phase transitions in CdTe and CsCl. 1 - endotherm peaks, 2 - secondary peak assigned to disassembling of melt structuring, 3 - exothermal peak, T_m - melting point, T_S - solidification point, $T_{S\text{-}S}$ - solid-solid transition*, ΔT - melt supercooling.

* Before melting at around 450°C the CsCl structure (α-CsCl) converts to the β-CsCl form with rock salt structure.

Fig. 2.3: Thermal effects at phase transitions in selected crystalline materials.

During a crystallization process, the latent heat of crystallization $\mathcal{L}$, released at the propagating liquid–solid interface, must be continuously transported away from it by using a well-balanced combination of both temperature gradient and crystallization velocity. For instance, when the heat balance is disturbed by a too high growth velocity, the temperature gradient at the crystallization front and, thus, the morphological interface stability can be affected markedly (see lecture part III). The *Spec box 2.1* shows the estimated quantity of heat produced at a propagating melt–solid interface during the growth of a cylindrical silicon crystal (at first approximation comparable with Czochralski growth).

Spec box 2.1: The heat quantity

Let us assume the crystallization of a cylindrical silicon boule with a constant diameter pulled from its melt toward the vertical direction *z*. The quantity of heat *dQ* produced by the crystallization of the solid volume amount *dV* due to temperature reduction *dT* is then

$$dQ = c_p \, \rho \, dV dT \tag{B2.1-1}$$

where c_p is the heat capacity and ρ the density of the solid phase. After expressing the crystallizing volume part $dV = (dz/dt)dF$ with time interval dt and area differential dF and replacing the heat capacity by the partial derivative of H at constant pressure $c_p = (dH/dT)_p$ becomes

$$dQ = \left(\frac{dH}{dT}\right)_p dT \, \rho \left(\frac{dz}{dt}\right) dF \, dt \tag{B2.1-2}$$

where $(dz/dt) = v_z$ is the crystallization velocity along the coordinate z, and dF the solidifying interface area. At constant cross section, it becomes

$$\frac{dQ}{dt} = \Delta H \, \rho \, v_z \, F \tag{B2.1-3}$$

For instance, during the growth of an 8-inch Czochralski silicon crystal with a constant cross section $F = \pi r^2 = 314$ cm^2 (r – crystal radius = 10 cm) crystallizing with a constant velocity $v_z = 10$ cm/h, after setting $\Delta H =$ 1.8 kJ/g and $\rho = 2.3$ g/cm^3 the continuously produced quantity of heat per second at the interface yields 3.61 kW.

Tab. 2.1: Abrupt change of thermodynamic quantities at first-order phase transition (*with permission from Trans. Tech, Publ.; supplemented values are from the public domain of Wikipedia*).

First-order phase transitions of selected materials						
Phase transition (1→2)	**Materials**	***T* and *p* at phase transition**		**Volume change $(V_1 - V_2)/V = \Delta V/V$**	**Entropy change $S_1 - S_2 = \Delta S$**	**Enthalpy change $H_1 - H_2 = \Delta H$**
		(K)	**(MPa)**	**(%)**	**(J/mol K)**	**(J/mol)**
Vapor → solid (V → S)	Al	723	0.1	>99.9	345	250×10^3
	Si	1,000	0.1	>99.9	304	304×10^3
Liquid → solid (V → S)	Al	933	0.1	−6.0	12	11×10^3
	Si	1,693	0.1	+9.6	30	50×10^3
	GaAs	1,511	0.2	+10.7	64	97×10^3
	ZnSe	1,733	0.4	−19.6	8	13×10^3
	$LiNbO_3$	1,533	10^{-8}	−20.9	17	26×10^3
Solid → solid (S → S)	$BaTiO_3$	393	(T → C)*	0	0.5	0.2×10^3
	ZnSe	1,698	(W → ZB)*	0	0.6	1.0×10^3
	CsCl	749	(α → β)*	−4	10	7.5×10^3
	C	1,200	(G → D)*	−37	1	1.2×10^3
Melt-solution → solid (ms → s)	*Solvent* CdTe	1,073	Te		43	46×10^3
	BaO	1,173	B_2O_3		18	21×10^3
	$Y_3Al_5O_{12}$	1,623	PbO/PbF_2*		15	25×10^3
	NaCl	300	H_2O		13	3.9×10^3

*T → C: tetragonal → cubic; W → ZB: wurtzite → zinc blende; α → β: α-CsC (CsCl structure) → β-CsC (NaCl structure); G → D: graphite → diamond; PbO/PbF_2: 1 PbO + 0.5 PbF_2.

Figure 2.4 shows the schematic curves of $H(T)$, $U(T)$, $TS(T)$, $pV(T)$, and $G(T)$ along the three phases of solid, liquid, and vapor, as well as the abrupt changes ΔH, $T\Delta S$, and $p\Delta V$ at the phase transition points at constant pressure p. Table 2.1 summarizes the values of abrupt changes of some system variables of selected materials at vapor–solid, liquid–solid, and solid–solid phase transitions. Some examples of solution–solid phase transitions are added. As can be seen, while the largest differences are occuring at the vapor–solid phase transition, but the smallest occur at solid–solid phase transitions.

Émile Clapeyron
(1799 – 1864)

Rudolf Clausius
(1822 – 1888)

slope of $p(T)$

$$\frac{dp}{dT} = \frac{\Delta S}{\Delta V}$$

vapour \rightarrow solid

$$\frac{d \ln p}{dT} \approx \frac{\Delta H}{RT^2}$$

see e.g. in:
K.-Th. Wilke, J. Bohm:
Kristallzüchtung
(Deutsch and Thun,
Frankfurt a.M. 1988)

Sketch of temperature courses of Gibbs potential G, thermodynamic variables H, U and energy terms TS and pV according Eqs. (2.1) and (2.2) at constant pressure. At the phase transitions T_{SL} and T_{LV} the variables are changing abruptly.

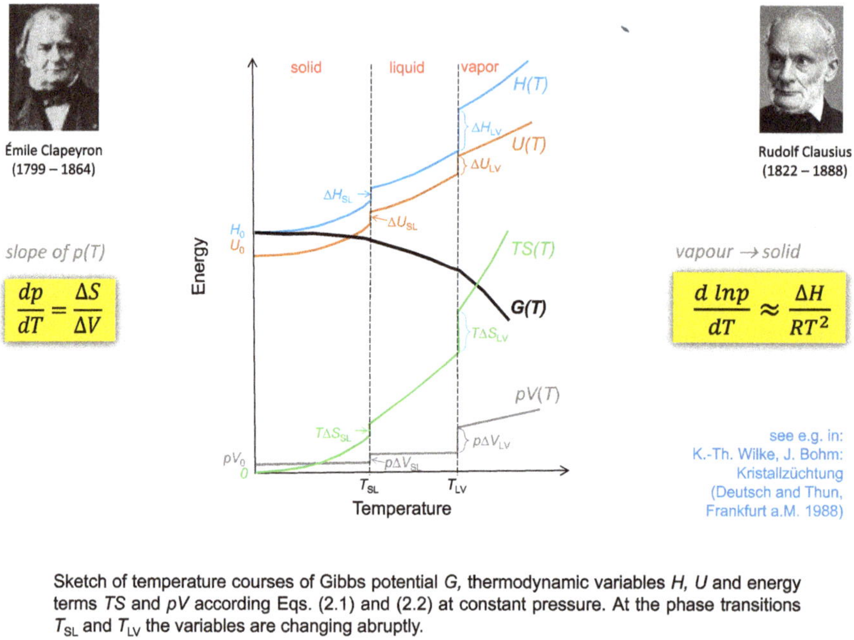

Fig. 2.4: Temperature profiles of thermodynamic quantities (*the portraits of É. Clapeyron and R. Clausius are in the public domain*).

During practical crystal growth processes with a first-order phase transition, each responsible grower can reconstruct the absorption and release of latent heat by observing the running heating power program and related temperature automation behavior. A particularly demonstrative example is the crystallization of huge silicon ingots for the production of solar cell wafers (Fig. 2.5). Both the heating and cooling curves, continuously checked by thermocouples, show the characteristic effect of thermal lingering due to the endothermic and exothermic reactions at the moments of melting and crystallization, respectively. As a result, the control unit commands a related increase or decrease of the electrical power.

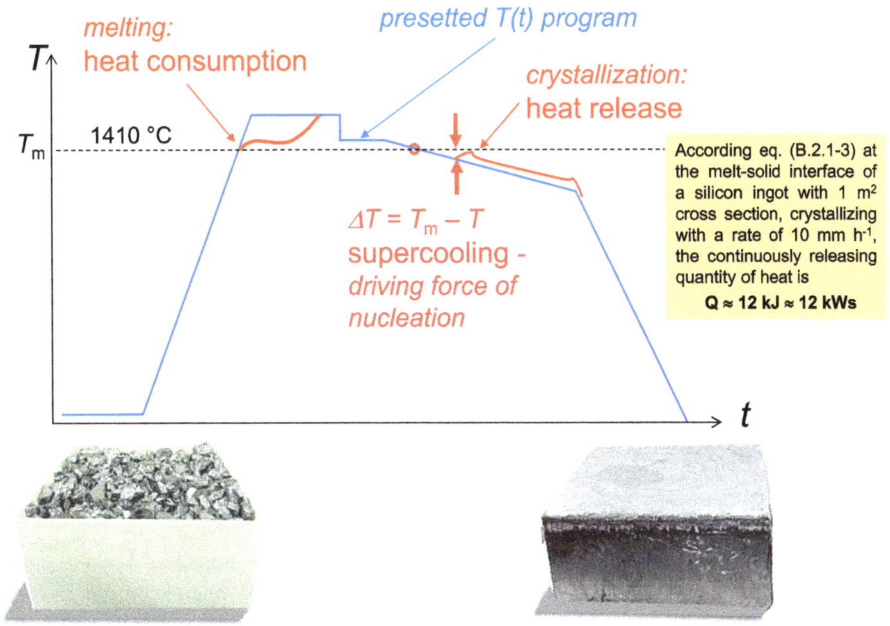
melting:
heat consumption
presetted T(t) program
crystallization:
heat release
T
Tm
1410 °C
ΔT = Tm – T
supercooling -
driving force of
nucleation
According eq. (B.2.1-3) at the melt-solid interface of a silicon ingot with 1 m² cross section, crystallizing with a rate of 10 mm h⁻¹, the continuously releasing quantity of heat is
Q ≈ 12 kJ ≈ 12 kWs
t

Fig. 2.5: Sketch of the temperature–time process during the melting and crystallization of a multicrystalline silicon ingot for solar cell production (*images of feedstock crucible and crystallized silicon ingot used with permission from IKZ Berlin*).

In the case of solution growth characterized by reduced crystallization temperature, the question arises as to how big the difference between the quantities of heat is. The estimation of the value of the heat of crystallization at growth from melt–solution is shown in *Spec box 2.2*. One of the components in a given compound system is used as a solvent in order to reduce the phase transition temperature. Figure 2.6 sketches the transition enthalpy at the melting point, ΔH_m, and at a melt–solution temperature, ΔH_{mS}, via a heating–cooling cycle. From the estimation in *Spec box 2.2*, it can be concluded that with minor differences between the heat capacities, there is no significant contrast between both enthalpies, and one can approximate $\Delta H_m \approx \Delta H_{mS}$.

Spec box 2.2: The enthalpy of solution

(*contributed by I. Pritula, V. Cherginets, ISC Kharkiv, Ukraine*)

According to the thermal cycle in Fig. 2.6, the enthalpy of crystallization from melt–solution is

$$\Delta H_{mS} = -(\Delta H_1 + \Delta H_m + \Delta H_2) \qquad \text{(B2.2-1)}$$

whereby the enthalpy change of a solid substance with heat capacity c_{ps} during the heating process 1 from a given melt-solution temperature T_{mS} to its melting point T_m can be expressed by

$$\Delta H_1 = \int_{T_{mS}}^{T_m} c_{ps}\, dT \approx c_{pS}(T_m - T_{mS}). \tag{B2.2-2}$$

After melting at T_m and release of ΔH_m, the liquid phase with heat capacity c_{pL} is cooled during process 2 down to the temperature T_{ms}, and the enthalpy interval is

$$\Delta H_2 = \int_{T_m}^{T_{mS}} c_{pL}\, dT \approx c_{pL}(T_{mS} - T_m). \tag{B2.2-3}$$

Setting $(T_m - T_{mS}) = \Delta T$ and inserting (B2.2-2) and (B2.2-3) into (B2.2-1), the released enthalpy of fusion at crystallization from melt-solution becomes

$$\Delta H_{ms} = -\left[\Delta H_m + \left(c_{pS} - c_{pL}\right)\Delta T\right] \tag{B2.2-4}$$

For example, let us estimate the value of ΔH_{ms} for a melt-solution crystal growth process of CdTe at $T_{mS} = 800$ °C (1,073 K), i.e., $\Delta T = (T_m - T_{mS}) = 1{,}365$ K – 1,073 K = 292 K below the melting point (T_m = 1,092 °C), which offers many benefits such as markedly reduced twin and dislocation densities. Such successful growth technology proves to be the traveling heater method, where CdTe crystals are grown from a Te-rich melt-solution zone. Taking $\Delta H_m = 209.2$ J/g, $c_{ps} = 0.2$ J/g K (at 800 °C), and $c_{pl} \approx 0.26$ J/g K (at 1,092 °C), the enthalpy of melt-solution-solid transformation is about −191.7 J/g = −46 kJ/mol yielding no significant difference to the latent heat of melting.

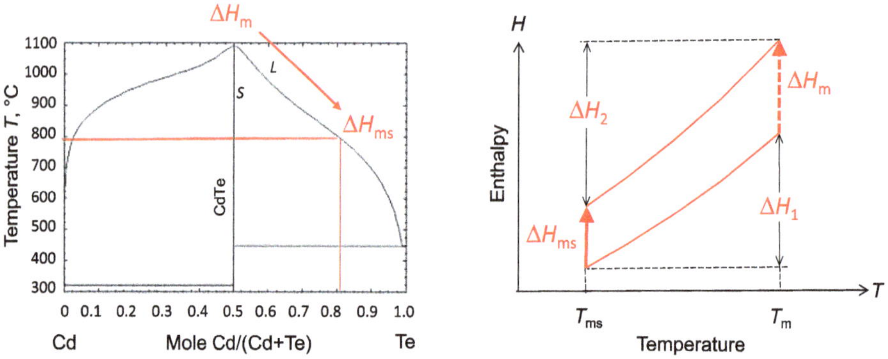

Left: T-x - projection of the system Cd-Te demonstrating the difference between ΔH_m at the melting point T_m and ΔH_{ms} at reduced temperature T_{ms} of melt-solution growth, e.g. by THM from a Te-rich zone at 800°C.

Right: scheme of thermal cycle for estimation the relation between enthalpy of melting and enthalpy of melt-solution (*see Spec box 2.2*).

Fig. 2.6: Enthalpy of solution (*left phase diagram with permission from Elsevier*).

Compared to a first-order phase transition, in a *second-order phase transition,* the system variables change continuously, and heat is neither released nor absorbed. Such a transition takes place within one phase and not between two phases. Examples of second-order phase transitions are the occurrence of a magnetic dipole moment in a magnetic substance upon transition from the paramagnetic to the ferromagnetic state or the appearance of superconductivity in metals and alloys, for example. Also, glass formation belongs to such a transition. Glass is a highly supercooled liquid-like state with all of the atoms or molecules unaligned with one another because it has been solidified at such a high viscosity that it prevents the transport of atoms or molecules from coming together and forming the well-structured solid phase. Figure 2.2 compares the sketched courses of enthalpy versus temperature for first- and second-order phase transitions.

3 Phase equilibrium and phase diagrams

3.1 Two phases of one component

Concerning eqs. (2.1) and (2.2), the description of the Gibbs free energy of one given phase is

$$G(T,p) = U + pV - TS = H - TS \tag{3.1}$$

The expression for the infinitesimal reversible change in the Gibbs free energy is then

$$dG = dH - d(TS) = dH - TdS - SdT \tag{3.2}$$

Adding the fundamental equation for the infinitesimal enthalpy change in dependency on the parameters S and p, the fundamental thermodynamic equation for Gibbs energy is obtained,

$$dH = TdS + Vdp \tag{3.3}$$

which becomes zero when infinitesimal fluctuations no longer effect any change:

$$dG = Vdp - SdT = 0 \tag{3.4}$$

If the variables T and p are constant, one obtains the slopes $\partial G/\partial p|_T = V$ and $\partial G/\partial T|_p = -S$, respectively, and the total differential of the Gibbs free energy is then

$$dG = \left.\frac{\partial G}{\partial p}\right|_T dp - \left.\frac{\partial G}{\partial T}\right|_p dT \tag{3.5}$$

From eqs. (3.4) and (3.5), it follows that in a closed system consisting of one phase $\phi = 1$ and one component $C = 1$ only, the temperature T and pressure p can be chosen independently in a wide range without phase transition. Expressed by the *Gibbs' phase rule*, this means that the system has two degrees of freedom f and exists, thus, under *two-variant conditions*

$$f = C - \phi + 2 = 2 \tag{3.6}$$

However, if T and p combination ranges to a point where the pure component undergoes a separation into two phases ($\phi = 2$), the freedom f decreases from 2 to 1 and it becomes no longer possible to control independently T and p, without phase transition, referred to as *invariant condition*. Two phases ($\phi = 2$) are in equilibrium when the Gibbs free energies of the phases (G_1 and G_2) are equal so that the potential difference between the phases becomes zero ($\Delta G = 0$). To illustrate the equilibrium between the two phases best, we cross the G–T–p planes of two phases within the three-dimensional phase space, as sketched in Fig. 3.1. The overlap of both planes results in

https://doi.org/10.1515/9783111711164-003

the equilibrium line within the two-dimensional p–T projection and the relevant equilibrium point in the G–T projection, where

$$G_1 - G_2 = \Delta G = \Delta U - T\Delta S + p\Delta V = 0 \quad (3.7)$$

Inserting $\Delta U + p\Delta V = \Delta H$ from eq. (3.1) becomes the *Helmholtz free energy* ΔF as

$$\Delta G|_{p,V} \cong \Delta F|_{p,V} = \Delta H - T\Delta S = 0 \quad (3.8)$$

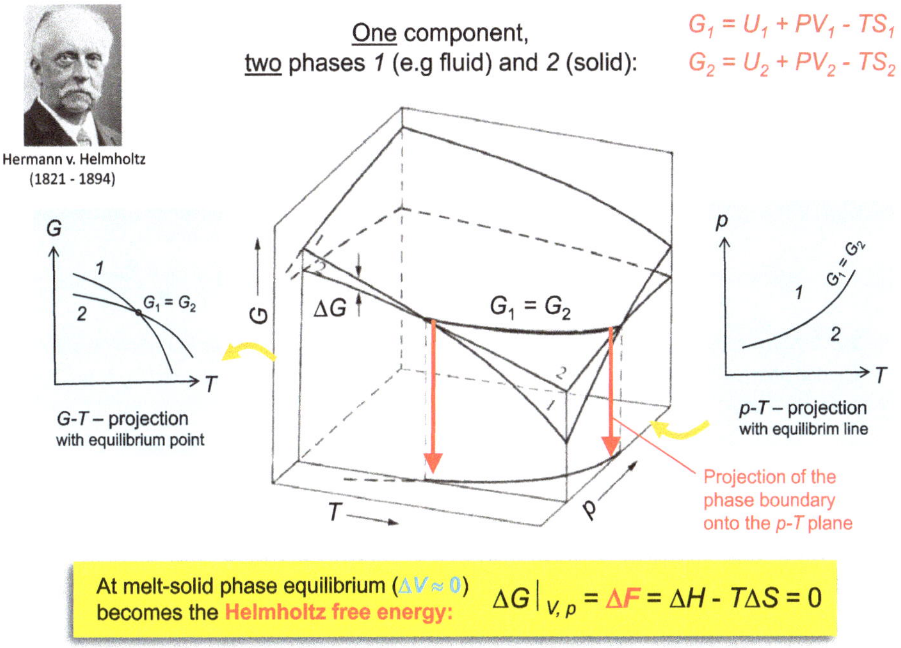

Fig. 3.1: Schematic *G*–*T*–*p* space with crossing phase planes 1 and 2 projecting phase boundaries on the *G*–*T* – and *p*–*T* – planes (*the portrait of H. Helmholtz is public domain*).

Figure 3.2 shows two examples of G–T projections of the monoatomic carbon and compound system GaN, treated as quasi one component. Due to the extreme melting conditions GaN cannot be grown from its stoichiometric melt but from its solving constituents under a given N_2 pressure. With increasing pressure, the equilibrium shifts to higher T and the GaN stability range is extended.

Figure 3.3 depicts a schematic three-dimensional phase diagram with its related two-dimensional p–T, T–V, and p–V projections of a one-component system (water might serve as an approximated model when it is treated as a quasi-one-component [H_2O]). As can be seen, the triple line between the three phases of vapor, liquid, and solid in the 3D diagram becomes triple point in the 2D p–T projection. Concerning the Gibbs' phase rule (eq. (3.6)) with $C = 1$ and $\phi = 3$, this represents an *invariant condition* where $f = 0$. Mostly, the p–T projection is used to demonstrate the crystallization (of

Gibbs free energy G vs. temperature T for solid graphite G_{SG}, liquid carbon G_L, and diamond G_{SD}.

adapted from J. Narayan et al., Mater. Res. Lett. 6 (2018) 353

Gibbs free energy G of solid GaN_S and its solving constituents liquid Ga_L and vaporous $\frac{1}{2}N_{2V}$ vs. temperature T. With increasing N_2 pressure the equilibrium point shifts to higher temperatures and the GaN stability range is extended.

adapted from S. Krukowski et al., Acta Physica Polonica B 37 (2006) 1265

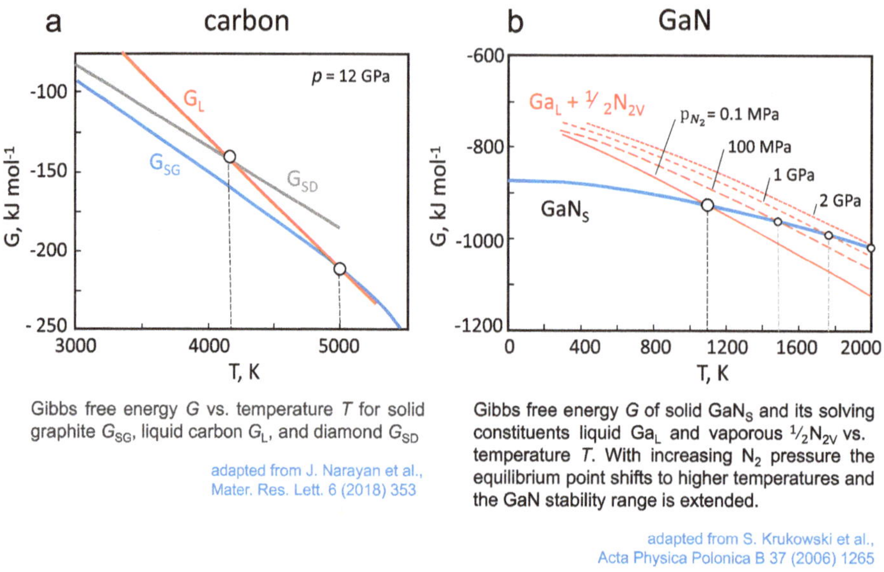

Fig. 3.2: *G–T projections with crossing equilibrium points of solid and liquid phases of carbon (a) and solid and solution phases of GaN (b) (open CCBY license of Taylor & Francis (a) and re-using according to the License CC-BY 4.0 (b)).*

Gibbs phase rule:

phase regions s, l, v
$$f = c - \phi + 2 = 1 - 1 + 2 = 2$$
divariant conditions

phase boundaries s-l, s-v, v-l:
$$f = 1 - 2 + 2 = 1$$
monovariant conditions

triple point/line (v-l-s):
$$f = 1 - 3 + 2 = 0$$
invariant conditions

see e.g. in: K.-Th. Wilke, J. Bohm: Kristallzüchtung (Deutsch and Thun, Frankfurt a.M. 1988)

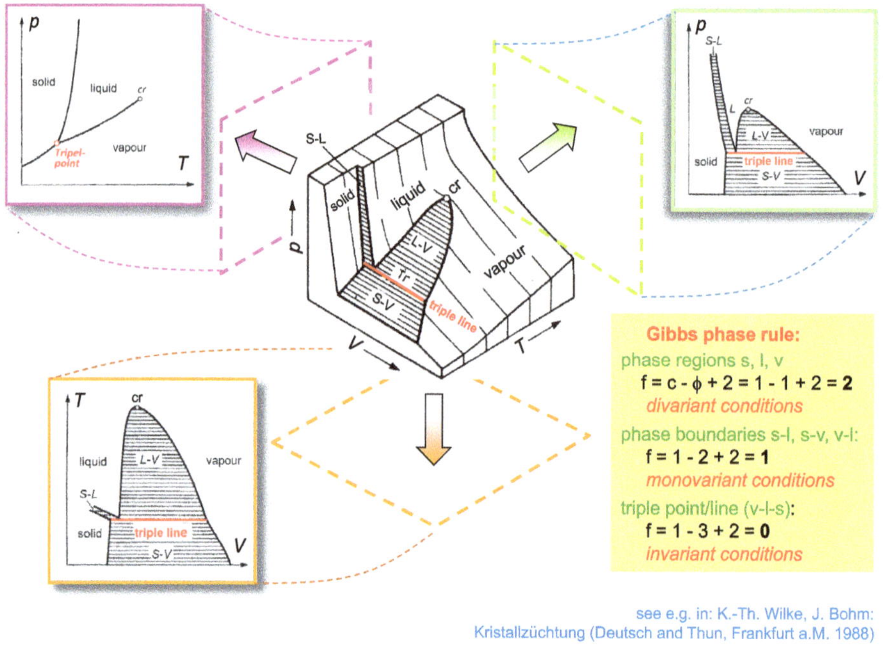

Fig. 3.3: Phase projections of a one-component p–T–V phase space with triple lines and points, illustrating the Gibbs phase rule (*the graphics are public domains of Wikipedia*).

ice) by either lowering T from liquid phase at moderate pressures p or by increasing p from gaseous phase at low temperatures below the triple point. The T–V and p–V projections are particularly suitable for demonstrating the volume differences between the phases and their changes within the phases. Figure 3.4 represents the p–T projections of the one-component systems, carbon with diamond phase and silicon.

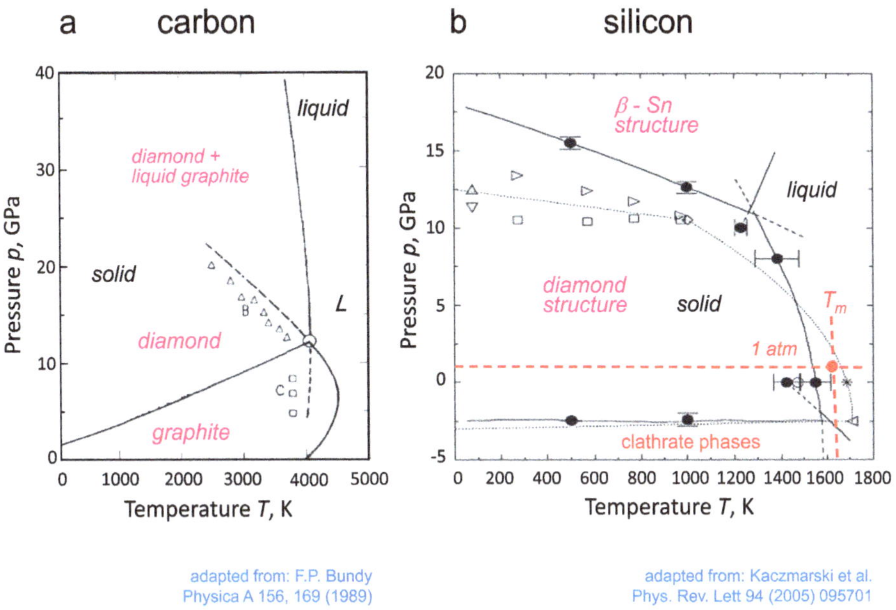

Fig. 3.4: p–T phase projections of one-component materials, such as carbon (a) and silicon (b) (*with permission of Elsevier (a) and APS (b)*).

It should be noted here that, correctly speaking, the designation "phase diagram" corresponds to the 3D representation only but not to its 2D "phase projections". Just a spatial image is able to show the phase planes via the consecutive thermodynamic parameters (also, the original Greek meaning διάγραμμα (*diágramma*) denoted a geometric shape). Therefore, in the following phase discussions, we prefer to use the term "projection" when a functional cut of only two variables is discussed. Of course, the colloquial language of phase diagram for a 2D projection should also be, furthermore, commonly used.

3.2 Two phases of two (or more) components

3.2.1 Basic principles

Now, several material (chemical) components are presented (in the following, specified by index $i = A, B, C, \ldots, K$). In dependency on the substance interaction mode, this can lead to totally or partially mixed systems, compound formation, and mutual insolubility (see below). *Spec box 3.1* shows some policies for the correct indication of some substance combinations (note, the specification by mole fraction is introduced in *Spec box 3.2*).

Therefore, the Gibbs free energy depends not only on the parameters T and p [see eq. (2.1)] but also on the component quantity n_i (substance mass or concentration of a given component i). Accordingly, the potential functionality is now

$$G = f(T, p, n_i) \tag{3.9}$$

In compliance with eq. (3.5), the total differential of the Gibbs free energy is then

$$dG = \left.\frac{\partial G}{\partial p}\right|_{T,n_i} dp - \left.\frac{\partial G}{\partial T}\right|_{p,n_i} dT + \sum_{i=A}^{K} \left.\frac{\partial G}{\partial n_i}\right|_{p,T} dn_i \tag{3.10}$$

with the partial derivatives $(\partial G/\partial p) = V$, $(\partial G/\partial T) = -S$ and the sum of potential dependencies on all chemical components $i = A, B, \ldots, K$ presented in a given system.

Spec box 3.1: The substance indication

Usually, a *binary completely mixed system* consisting of two substances A and B is symbolized by A–B (with a hyphen), like Si–Ge.

A *ternary mixed system* is characterized by A–B–C such as Ga–Al–As.

A *compound* is usually typified by AB (without hyphen), like GaAs or ABC_m, like $LiNbO_3$.

The addition of a *dopant* B, C, to a single matrix substance A is designated as A:B,C, . . . (via a colon), like Si:As. In the case of compounds AB, the marking of added atoms should start with C as AB: C, like SiC:Al. On the other hand, a compound can be also thought as "quasi-one-component A" so that the dopant terms can start with B as (AB ≡ A):B, like GaN (A):Mg (B).

Note that often, especially in the crystal growth of laser materials, the doping element is placed before the matrix component in order to express its incorporation into the crystal via a colon, e.g., Yb: YAG or Fe:ZnSe.

Normally, pro rata components deviating as *excess from stoichiometry* in a compound AB are referred to A or B and this state is indicated as AB–A or AB–B, like CdTe–Cd or CdTe–Te.

Material systems of *mutually insoluble* substances are written with a separating "+" between the parts, i.e., A + B, AB + AC, $ABC_m + BC_n$, as is well known from eutectic compositions such as Pb + Sn, SnSe + ZnSe, $PbMoO_4 + Mo_3$, respectively.

From eq. (3.10) follows that, at constant T and p, the change of the Gibbs potential would depend on the variation of n_i only. The related partial derivative is named *partial or molar Gibbs free potential* g, to be equated with the *chemical potential* μ_i

$$\left.\frac{\partial G}{\partial n_i}\right|_{T,p} = g = \mu_i \tag{3.11}$$

The derivation of G over n_i translates the extensive value of the potential to an intensive one. As is well known, an *intensive property* does not depend on the system size or the amount of material in the system. Examples of further intensive properties are the temperature T, pressure p, heat capacity c_p, density ρ, and concentration c. Thus, treatments with intensive chemical potential μ_i require the intensification of the other involved variables too, i.e.,

$$\frac{V}{n_i} = \text{v (specific volume)}; \quad \frac{H}{n_i} = h \text{ (specific enthalpy)}; \quad \frac{S}{n_i} = s \text{ (specific entropy)}. \quad (3.12)$$

In order to express the component proportion in material systems with substituting miscibility, such as mixed crystals Si–Ge or CdTe–ZnTe, or solid-solutions between crystalline matrix and dopants or impurities, such as Si:As or Al_2O_3:Cr, one uses the dimensionless *mole fraction* x_i

$$\frac{n_A}{n_A + n_B} = x_A = (1 - x_B) \quad (3.13)$$

$$\frac{n_B}{n_A + n_B} = x_B = (1 - x_A) \quad (3.14)$$

Multiplying the mole fraction by 100 gives the *mole percentage*, also referred to the as amount percent. The *mass fraction* w_i is calculated by the formula

$$w_i = x_i \frac{M_i}{M} \quad (3.15)$$

where M_i is the molar mass of the i-component and M the average molar mass of the mixture.

In addition to binary material compounds that contain atoms of two given groups of the periodic table at the same proportions, e.g., III–V, II–VI, I–VII, or multiple material compounds such as II–IV–V_m, mixed forms can also be produced within the groups, in which the proportion of atoms in one of the groups, e.g., III or V is composed of two atomic varieties. This creates *pseudo-binary* or *ternary* (three atomic varieties in total) mixed systems and *pseudo-ternary or quaternary* (four atomic varieties in total) mixed systems. *Spec box 3.2* summarizes the correct handling of mole fraction for the related material systems.

After the mole fraction has been introduced, the molar Gibbs free energy for material systems, consisting of several components i, can be expressed in the summary form. Taking n as the sum of all material quantities at fixed T and p, the total molar Gibbs free energy is

$$g = \left.\frac{G}{n}\right|_{T,p} = \sum_{i=A}^{K} \mu_i \frac{n_i}{n} = \sum_{i=A}^{K} \mu_i x_i \quad (3.16)$$

named *Euler's theorem*. Thus, the chemical potential μ_i is concentration-dependent. To apply it to specific phases, one must connect it to specific equations of state in order to define their concentration and temperature dependence. For that, one can use the ideal gas equation which says that any reaction involving liquid and solid phases is described

Spec box 3.2: The mole fraction in systems with compounds

The indication for *binary mixed crystals* A–B and *pseudo-binary ternary mixed crystals* AB–AC or AB–CB via mole fraction x_i is $A_{1-x}B_x$ and $AB_{1-x}C_x$ or $A_{1-x}C_xB$, respectively. For example, Si-Ge, GaAs-GaP, and CdTe–ZnTe are written as $Si_{1-x}Ge_x$, $GaAs_{1-x}P_x$ and $Cd_{1-x}Zn_xTe$, respectively.

For a *pseudo-ternary* or *quaternary mixed system*, two mole fractions x_i and y_i are required to indicate the component relation $A_{1-x}B_xC_{1-y}D_y$ with $x = x_A = n_A/(n_A + n_B)$ and $y = y_C = n_C/(n_C + n_D)$ or $AB_{1-x-y}C_xD_y$ with $x = x_B = n_B/(n_B + n_C)$ and $y = y_B = n_B/(n_B + n_D)$. Such examples are well known from epitaxy of optoelectronic devices such as $Ga_xIn_{1-x}As_yP_{1-y}$ or $GaAs_{1-x-y}Sb_xN_y$, respectively.

At this point, it is important to note that in the literature for mixed systems, very often a chemically incorrect indication is used, whereby they are presented as compounds like "SiGe", "CdZnTe", "GaAlN", "GaAsSbN", etc., i.e., by intentionally omitting the mole fraction as indices. Against this, would be the acceptable use of round brackets like Si(Ge), (Cd,Zn)Te, (Ga,In)(As,P), Ga(As,Sb,N), a.s.o., when for convenience, one likes to neglect the index x.

Next frequently asked question is which scale of mole fraction is favorable to use for a binary system A–B with compound AB formation? In other words, which value of x_i is best manageable to indicate the component ratio between the compound AB and a certain excess of one of the components A or B? For instance, this is the case at melt–solution growth when one compound partner A or B is applied as solvent. The following two scaling variants are possible:

i) x_i is used for the determination of the total A/B ratio as $n_A/(n_A + n_B)$, i.e., from pure component A with $x_B = 0$ (≡ 100% A: 0% B) to pure B with $x_B = 1$ (≡ 0% A: 100% B) so that the compound AB is indicated by the mole fraction $x_B = 0.5$. This variant is favorable for weighing a feedstock before the synthesis of AB is started, for example.

ii) x_i is used for the region between the compound AB and one of the substance partners A or B only. Taking for this case, the index $y_B = n_B/(n_B + n_{AB}) = 1$, for 0% AB: 100% B and $y_B = 0$ for 100% AB: 0% B. Such a variant is favorable for use at melt–solution growth of a compound AB, starting from a seed crystal AB in a solvent A or B. The relation between both mole fractions is sketched as following.

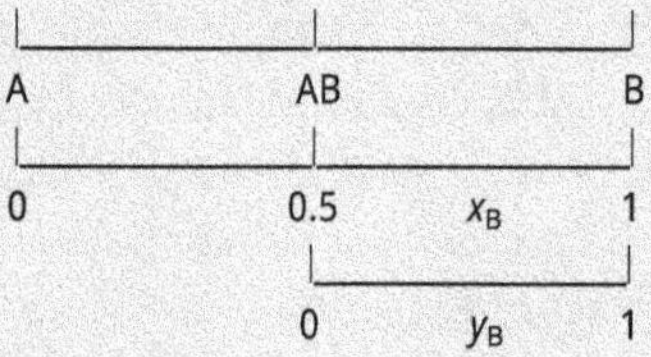

Because

$$x_B = 1\big|_{y_B=1} \quad \text{and} \quad x_B = 0.5\big|_{y_B=0} \tag{B3.2-1}$$

within the region $0.5 \le x_B \le 1$, it applies

$$\frac{(1 - x_B)}{(1 - y_B)} = \frac{x_B}{1} \tag{B3.2-2}$$

and after simply rearranging eq. (B3.2-2), the value of y_B becomes

$$y_B = \frac{(2x_B - 1)}{x_B} = 2 - \frac{1}{x_B} \tag{B3.2-3}$$

by reaction between vapors of the substances (see *Spec box 3.3*). Accordingly, the chemical potential of a component i in a particular phase of solution is

$$\mu_i = \mu_i^0(T, p) + RT \ln (\gamma_i x_i) \tag{3.17}$$

where μ_i^0 is the standard potential (chemical potential of pure species i), R the universal gas constant, and γ_i the *activity coefficient* of a given component i. The value γ_i is the coupling factor between the mole fraction of a given substance and its *activity* a_i, also named *fugacity* in gas-related expressions:

$$a_i = \gamma_i \, x_i \tag{3.18}$$

Thus, the activity is used to account for deviations from ideal behavior in a mixture of substances. For instance, in an ideal mixture (ideal solution), the microscopic interactions between each pair of chemical species are the same (or macroscopically equivalent, i.e., the enthalpy change of solution and volume variation in mixing is zero) and, therefore, the activity coefficient γ_i becomes unity. In this case, the properties of the solution can be expressed directly in terms of simple concentrations or partial pressures of the substances according to the *Raoult's law*. The deviations from ideality, however, are expressed by $\gamma_i \neq 1$ (see below).

Generally, in a given material mixture containing both the ideal and real parts of chemical potential, one can combine as a sum of an ideal contribution and an excess contribution

$$\mu_i = \mu_i^{\mathrm{id}} + \mu_i^{\mathrm{ex}} \tag{3.19}$$

whereas the ideal part depends on the mole fraction x_i only. But the intermolecular interaction between the same species and with other presented species, expressed by the activity according to eq. (3.18), is considered in the excess part.

Now, let us look at the component correlation in the energetic consensus. In addition to the standard molar free energy of the presented pure components g_0, the total molar free energy $g = h - Ts$ contains a mixing term Δg_m

$$g = g_0 + \Delta g_m \tag{3.20}$$

According to eqs. (3.16) and (3.19), the functional expression via chemical potentials and mole fraction is

$$g(x) = \sum_{i=A}^{Z} \mu_i^0 x_i + \Delta\mu_m \tag{3.21}$$

where μ_i^0 is the standard chemical potential of the presented pure components and $\Delta\mu_m$ the mixed potential. Referring to eq. (3.1), it becomes

$$\mu(x) = \sum_{i=A}^{Z} \mu_i^0 x_i + (\Delta h_m - T\Delta s_m) \tag{3.22}$$

where Δh_m and Δs_m are the molar *enthalpy of mixing* and molar *entropy of mixing*, respectively [see eqs. (3.24) and (3.25)].

As is well known, the enthalpy of mixing (or heat of mixing or excess enthalpy) is the energetic quantity liberated or absorbed from a substance upon mixing. When a substance or compound is combined with any other substance or compound, the enthalpy of mixing is the consequence of the new interactions between the two substances or compounds. If the mixture proceeds at constant pressure, the mixing enthalpy is the same as mixing heat. The total molar enthalpy of the system, after mixing, is then

$$h_{\text{mixture}} = \Delta h_m + \sum x_i h_i \tag{3.23}$$

where h_i is the standard enthalpy of pure components i.

Spec box 3.3: The chemical potential of ideal and real solution

In an *ideal gas*, the specific (molar) volume v_{mol} at constant temperature T is given by the first derivative of the chemical potential μ from the pressure p, which is coupled with the ideal gas law as

$$V_{\text{mol}} = \left.\frac{\partial \mu}{\partial p}\right|_T = \frac{RT}{p} \tag{B3.3-1}$$

from which follows

$$d\mu = \int_{p_0}^{p} RT\frac{dp}{p} = RT\, d\ln\frac{p}{p_0}. \tag{B3.3-2}$$

Consequently,

$$\mu = \mu^0(T) + RT\ln\frac{p}{p_0} \tag{B3.3-3}$$

and when p_0 is chosen as unity (1 atm), one obtains the convenient form

$$\mu = \mu^0(T) + RT\ln p. \tag{B3.3-4}$$

where $\mu^0(T)$ is the standard chemical potential that depends on the temperature only.

In an *ideal solid mixture*, each partner acts like an ideal gas. Thus, eq. (B3.3-4) must also be valid for each individual partner i with partial pressure p_i as

$$\mu_i = \mu_i^0(T) + RT\ln p_i \tag{B3.3-5}$$

According to *Raoult's law*, the partial pressure of a substance over an ideal solution equals the product of the mole fraction and the total pressure,

$$p_i = x_i\, p \tag{B3.3-6}$$

The insertion of eq. (B3.3-6) into eq. (B3.3-5) yields the following partial chemical potential:

$$\mu_i = \mu_i^0(T) + RT\ln p + RT\ln x_i \tag{B3.3-7}$$

When using the mole fraction as measure of concentration, the first and second terms on the right side are combined as standard potential, depending on T and p, and the partial chemical potential in *ideal solutions* is then

$$\mu_i = \mu_i^0(T,p) + RT \ln x_i \tag{B3.3-8}$$

On the other hand, for *real solutions*, we have to use eq. (3.18), and the partial chemical potential relates to the activity as

$$\mu_i = \mu_i^0(T,p) + RT \ln a_i \tag{B3.3-9}$$

In other words, for an ideal solution, the activity coefficient of the mixed materials is $\gamma_i = 1$ and the activity a_i equals the mole fraction x_i as is used in the second term of eq. (B3.3-9).

Usually, one has to consider that the *chemical potential depends upon pressure and temperature.* It is convenient to approximate this dependency by using the linear functions:

$$\mu_i(T) = \mu_i(T_0) + \alpha(T - T_0) \tag{B3.3-10}$$

$$\mu_i(p) = \mu_i(p_0) + \beta(p - p_0) \tag{B3.3-11}$$

Here, α is the temperature coefficient and β is the pressure coefficient of the chemical potential. The equations allow us to calculate the chemical potential at temperature T and pressure p, if the potentials $\mu_i(T_0)$ and $\mu_i(p_0)$ at temperature T_0 and pressure p_0 are known. The coefficients α and β are given by the derivatives $\alpha = (\partial\mu_i/\partial T)_{p,x} = -s$ and $\beta = (\partial\mu_i/\partial p)_{T,x} = v$, where s is the molar entropy and v is the molar volume.

The molar mixing enthalpy Δh_m consists of the partial molar mixing (excess) enthalpies of each component Δh_{im}, where $i = A, B, \ldots, K$

$$\Delta h_m = x_A \Delta h_{Am} + x_B \Delta h_{Bm} + \cdots + x_K \Delta h_{Km} = \sum_{i=A}^{K} x_i \Delta h_{im} \tag{3.24}$$

The entropy of mixing (or configurational entropy) provides information about the constitutive differences of intermolecular forces or specific molecular effects in the materials. For mixed crystals, comparable with many-particle systems, the statistical concept of randomness can be used. Accordingly, the ideal mixing of materials is regarded as random at the atomic or molecular level. This is qualitatively easily visualized in terms of the increased disorder brought about by mixing.

On the other hand, mixing of nonideal materials may be nonrandom or rather ordered, exhibiting a reduced value of mixing entropy. In general, the molar entropy of mixing of i components with mole fractions x_i is

$$\Delta s_m = -R \sum_{i=A}^{K} x_i \ln x_i \tag{3.25}$$

with R the gas constant. The derivation of eq. (3.25) is shown in *Spec box 3.4.*

Spec box 3.4: The entropy of mixing

When two pure substances *A* and *B* are mixed under normal conditions, there is usually an increase of the system entropy. The comparison with the mixing of ideal gases is then obvious. Since the molecules of ideal gases do not interact, the increase in entropy must simply result from the extra volume available to each gas, on mixing. Thus, for gas *A* the available molar volume has increased from v_A to $(v_A + v_B)$. By calculating the entropy of expansion of each gas, we can conclude on the entropy of mixing as the following.

From the first and second law of thermodynamics, we can write

$$dU = TdS - p\, d\text{V} \tag{B3.4-1}$$

and after transposition, we obtain

$$dS = \frac{dU}{T} + \frac{p}{T} d\text{V} \tag{B3.4-2}$$

Because for ideal gases $dU = 0$ and $pV = n_i RT$, the entropy of expansion of the gas *A* into the total volume *V* of both *A* and *B* is obtained from

$$\int_{S_A}^{S_B} dS = n_i R \int_{\text{V}_A}^{\text{V}_B} \frac{d\text{V}}{\text{V}} = n_i R \int_{\text{V}_A}^{\text{V}_B} d\ \ln \text{v} \tag{B3.4-3}$$

so that the entropy of expansion of each gas *A* and *B* becomes

$$\Delta s_A = -R \ln\left(\frac{\text{V}_A + \text{V}_B}{\text{V}_A}\right) \text{ and } \Delta s_B = -R \ln\left(\frac{\text{V}_A + \text{V}_B}{\text{V}_B}\right) \tag{B3.4-4}$$

and because $V_A = n_A V_{Am}$, $V_B = n_B V_{Bm}$ with n_A, n_B, the component quantities, and $V_{Am} = V_{Bm} = V_m$, the mixing volume at equal p_0,T for gases *A* and *B* becomes

$$\frac{\text{V}_A + \text{V}_B}{\text{V}_A} = \frac{n_A \text{V}_m + n_B \text{V}_m}{n_A \text{V}_m} = \frac{n_A + n_B}{n_A} = \frac{1}{x_A} \text{ and } \frac{n_A + n_B}{n_B} = \frac{1}{x_B} \tag{B3.4-5}$$

with $\ln \frac{1}{x_i} = -\ln x_i$ it becomes $\Delta s_m = -R(n_A \ln n_A + n_B \ln n_B)$. Dividing it by the total component quantity $(n_A + n_B)$, the molar entropy of mixing is

$$\Delta s_m = -R(x_A \ln x_A + x_B \ln x_B) \tag{B3.4-6}$$

Expanding this equation to a multicomponent system becomes eq. (3.25).

Note, during a process of crystallization also, the presented atoms or molecules of differing components do mix either immediately at the growing solid-fluid interface or within the cooling crystalline phase by interdiffusion. As a result, solid mixtures of two or more components are formed as mixed crystal or alloy. Again, the same equations for the entropy of mixing can be used, but only for homogeneous uniform phases.

3.2.2 Ideal mixed systems

We start with the molar Gibbs free energy of mixing ($\equiv$ chemical potential of mixing)

$$\Delta g_m = \Delta h_m - T\Delta s_m \tag{3.26}$$

with Δh_m and Δs_m the molar mixing enthalpy and mixing entropy, respectively. A material system with ideal mixing is an ideal solid or fluid solution with thermodynamic properties, analogous to those of a mixture of ideal gases. That means, there are quasi no intermolecular interactions between the species of the constituting components, and the ideal mixing enthalpy becomes zero:

$$\Delta h_m^{\mathrm{id}} = 0 \tag{3.27}$$

Therefore, the activity coefficient γ_i, indicated by eq. (3.18), is equal to 1 and, thus, the activity a_i is expressed by the mole fraction x_i as

$$a_i = x_i \tag{3.28}$$

and the ideal partial chemical potential is

$$\mu_i^{\mathrm{id}} = \mu_i^0(T,p) + RT \ln x_i \tag{3.29}$$

As compared to the zero-enthalpy of mixing, the entropy of mixing is still acting to promote disordering during mixing. As a result, the configurational entropy is increasing versus mole fraction, as given in eq. (3.25). Assuming that the number of the components participating in the mixing process is only two, namely A the crystalline matrix and B dissolved in it. The molar entropy of mixing in eq. (3.25) becomes, for such a binary ideal A–B mixed system (in our case mixed crystal),

$$\Delta s_m^{\mathrm{id}} = -R(x_A \ln x_A + x_B \ln x_B) \tag{3.30}$$

and according to eqs. (3.22) and (3.29), the total chemical potential is modified as follows:

$$\mu(x)^{\mathrm{id}} = (\mu_A^0 x_A + \mu_B^0 x_B) - T\Delta s_m^{\mathrm{id}} \tag{3.31}$$

$$\mu(x)^{\mathrm{id}} = \underbrace{(\mu_A^0 x_A + \mu_B^0 x_B)} + RT\,\underbrace{(x_A \ln x_A + x_B \ln x_B)} \tag{3.32}$$

$$\mu(x)^{\mathrm{id}} = \qquad \mu_i^0(x) \qquad + \qquad \Delta\mu_m^{\mathrm{id}}(x) \tag{3.33}$$

where $\mu_A{}^0$ and $\mu_B{}^0$ are the standard chemical potentials of pure components A and B, respectively, $\mu_i{}^0(x)$ is the total standard potential, and $\Delta\mu_m{}^{\mathrm{id}}$ the relative chemical potential difference between the chemical potential μ_i of a component in solution and the chemical potential $\mu_i{}^0$ of the same component in a standard state. If we just consider only the chemical potential of an ideal mixing, we will obtain,

$$\Delta\mu_m^{\mathrm{id}}(x) = -T\Delta s_m^{\mathrm{id}} = RT(x_A \ln x_A + x_B \ln x_B) \tag{3.34}$$

Figure 3.5 shows the total chemical potential μ, as function of the mole fraction x_B for a given temperature T in an ideal mixed system of one phase (e.g., solid), according to eq. (3.34). As we can see, the total chemical potential $\mu(x)^{\mathrm{id}}$ forms a catenary curve below the straight line of the total standard potential $\mu^0(x)$, whereupon, the degree of its bending is determined by the potential of ideal mixing $\Delta\mu_m^{id}(x) = -T\Delta s_m^{id}$. The minimum chemical potential is found at mole fraction 0.5 when both components are completely mixed in each other in equal proportions. Note, considering the proportionality $d\mu \sim -s\,dT$, the curvature of the total chemical potential $\mu(x)$ is decreasing with increasing temperature.

$$\Delta\mu_m^{id}(x) = -T\Delta s_m^{id} = RT(x_A \ln x_A + x_B \ln x_B)$$

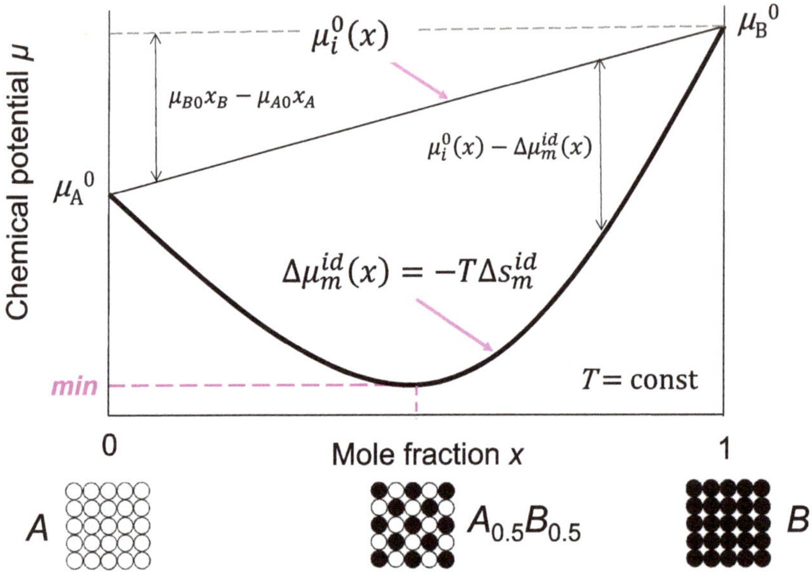

Fig. 3.5: Total chemical potential versus mole fraction in an ideal mixed phase.

Now let us consider the presence of *two ideally mixed phases* like a *liquid* (melt) and a *solid* (crystal), each consisting of two components A and B with mole fraction x_B that increases with content of B. Both states have a total chemical potential $\mu_L(x)$ and $\mu_S(x)$ as is shown by the $\mu(x)$ curves at four different temperatures in Fig. 3.6a. At high temperature T_6, the potential for the liquid solution is lower than that of the solid phase over the whole composition range and, therefore, the stable state. In contrast, at low temperature T_1, the potential for the solid solution is lower than that of the liquid phase. At certain temperatures, in the range $T_5 < T < T_2$, the two functions $\mu_L(x)$ and $\mu_S(x)$ are intersecting. Their contact starts at temperature T_5, identical with the melting point of the higher-melting component A and ends at T_2, the low melting point of component B. At both melting points, liquids A_L and B_L and solids B_S and A_S are in equilibrium. In between, the stable state with the lowest chemical potential is a mixture of solid and liquid solutions, defined by the contact points of a common tangent (see $\mu(x)$ projections at T_3 and T_4 in Fig. 3.6a). Hence, the compositions of the coexisting phases are those at the points of tangency (the tangent construction is given in the *Spec box 3.5*). Applying the potential situation for each temperature step, one can construct the T–x phase projection for an ideal mixed system $A_{1-x}B_x$, sketched in Fig. 3.6b. The distance between liquidus (S) and solidus (L) is determined by the proximity of the curves of the mixing entropy in both phases. The mole relation between S and L defines the thermodynamic equilibrium coefficient k_0, which will be discussed in more detail in Section 3.2.6.

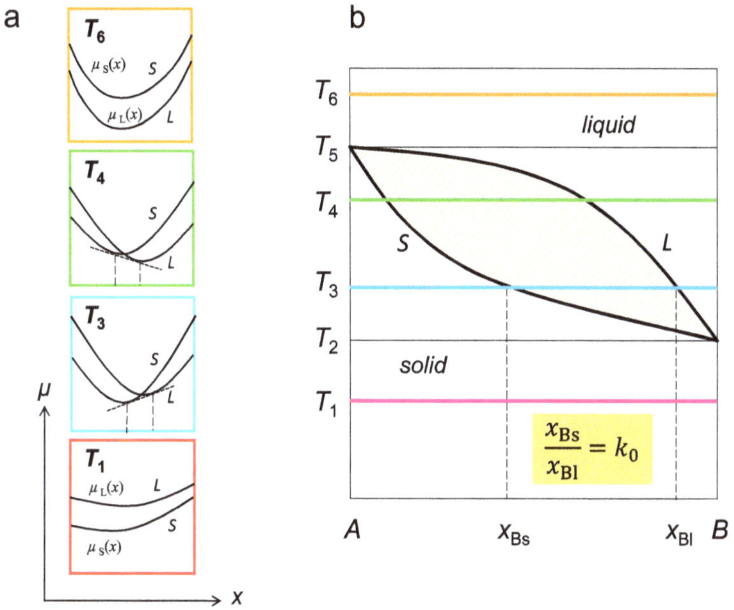

Fig. 3.6: μ–x (a) and T–x projections (b) of a system with near ideal mixing in the liquid and solid phases (k_0 – equilibrium segregation coefficient).

Concerning the *tangent construction* sketched in Fig. 3.7, it often arises the question how can it be that the total free energy between the minima of both the catenary curves, $\mu_S(x_B)$ and $\mu_L(x_B)$, and within the concentration region, $x_{BS}^{\min} < x_B < x_{BL}^{\min}$ (see Fig. 3.7), be lower than the related curve values, but corresponds with the tangent line? The answer is: although the potentials of the $\mu_{S,L}(x_B)$ curves are differing within this region, it does not yet mean that the system is in equilibrium when one of the curves is below the other; it reaches its minimum free enthalpy in the form of a *two-phase state* $S + L$, consisting at a given T of the concentration ratio $x_{BS}^{\min}/x_{BL}^{\min}$, i.e., they both touch the points of the tangent. In other words, if the A and B atoms in the homogenous liquid solution rearrange, a portion transforms to a solid, with composition Δx_{BS}, and a portion remains in a liquid solution, with composition altered to Δx_{BL}; the heterogeneous solid-liquid mixture takes on the molar free energy $\bar{g}^{eq} \equiv \bar{\mu}^{eq}$, being lower than that of the homogeneous liquid solution (within the range of $x_{BS}^{min} < x_B < x_{BL}^{min}$). The geometrical calculation proof, according to Fig. 3.7, is given in *Spec box 3.5*.

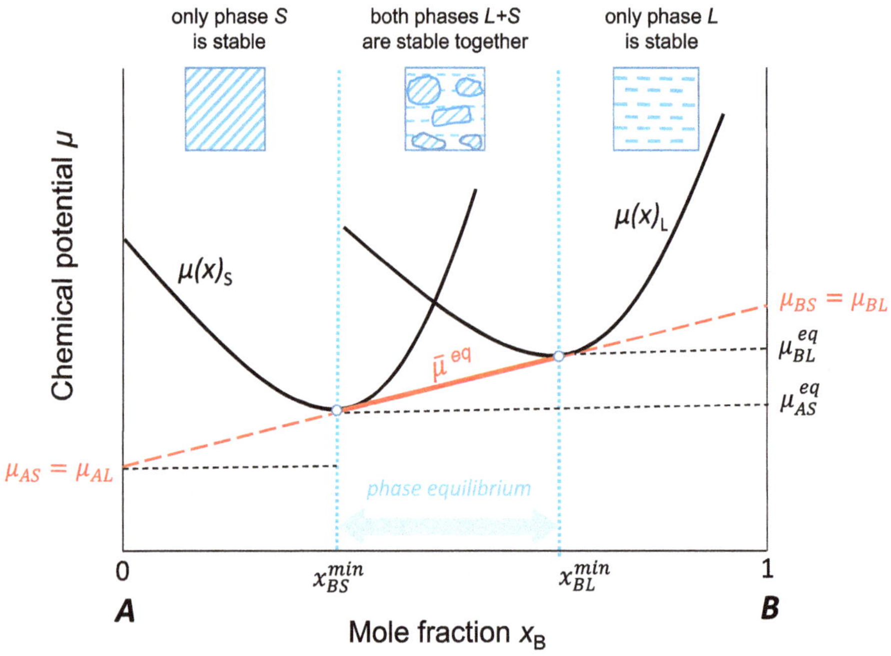

Fig. 3.7: Two-phase equilibration in a binary system *A–B*, demonstrated by the tangent construction.

Spec box 3.5: The tangent construction

The total chemical potential of a two-component system *A–B* is

$$\mu = \mu_A x + \mu_B(1-x) \tag{B3.5-1}$$

From Fig. 3.7, it follows that phase *S* has, at the mole fraction x_{BS}, the potential $\mu_S(x_{BS})$, and phase *L* has, at the mole fraction x_{BL}, the potential $\mu_L(x_{BL})$. The equilibrium potential μ_{eq} along the tangent between the two curves $\mu(x)_S$ and $\mu(x)_L$ is given by both slopes at the points x_{BS}^{min} and x_{BL}^{min} as

$$\frac{\partial \mu_S}{\partial x} = \mu_{AS}^{eq} - \mu_{AS} \quad \text{and} \quad \frac{\partial \mu_L}{\partial x} = \mu_{BL} - \mu_{BL}^{eq} \tag{B3.5-2}$$

Generally, two phases (e.g., solid and liquid) are in equilibrium when the chemical potentials of all constituents are identical. In a binary *A–B* system, it follows that

$$\mu_{AS} = \mu_{AL} \quad \text{and} \quad \mu_{BS} = \mu_{BL} \tag{B3.5-3}$$

The convertibility of chemical potentials from eq. (B3.5-3) in eq. (B3.5-2) shows the equality of both derivatives, i.e., slopes at both tangential contact points of the $\mu_L(x)$ and $\mu_S(x)$ curves. As a consequence, *the common tangent construction ensures the equality of chemical potentials of A and B in the solid and liquid phases* within the range $x_{BS}^{min} < x_B < x_{BL}^{min}$. Thus, it proves to be the criterion for phase equilibrium and, simultaneously, minimizes the total Gibbs energy by the equality of the chemical potentials. The chemical potentials that are in equilibrium at the tangential contact points are μ_{AS}^{eq}, μ_{BS}^{eq}, μ_{AS}^{eq}, and μ_{BL}^{eq}.

According to eq. (B3.5-1) and the equality of chemical potentials in the liquid and solid phases, the total chemical potential in the range $x_{BS}^{min} < x_B < x_{BL}^{min}$ is given by

$$\bar{\mu}^{eq}(x) = \mu_{AS}^{eq}x + \mu_{BL}^{eq}(1-x) = \mu_{AL}^{eq}x + \mu_{BS}^{eq}(1-x) \quad \text{(B3.5-4)}$$

For instance, setting $x = 1$ and 0 becomes $\mu_{AS}^{eq} = \mu_{AL}^{eq}$ and $\mu_{BL}^{eq} = \mu_{BS}^{eq}$, respectively. This means that the chemical potentials in both *S* and *L* phases are equal due to the tangential determination of the equilibrium contact points $x_{BS}^{min} = x_B^{eq}$ and $x_{BL}^{min} = x_{(1-A)L}^{min} = x_{(1-A)}^{eq}$.

There remains the question: Are there such material systems of ideal mixing over the whole composition range? Principally, such a state must be subject to strict conditions. In 1932, *Hume-Rothery* postulated the following rules for the ideal mixing of two elements: i) their atomic radii must be within about 15% of each other, ii) they must have the same crystal structure, iii) they must have similar electronegativity values, and iv) they must have the same valence, which is a measure of an atom's ability to combine with other atoms. That means, only isomorph components are able to realize such a complete intersolubility.

Actually, one can find several ideal mixtures between liquids, such as methanol and ethanol or CCl_4–$SnCl_4$, for example. In contrast, the finding of ideal solid solutions is very limited. Examples are the phase diagrams of the mixed crystal systems Ag–Au (Fig. 3.8a) and HgTe–HgSe, which show a near convergence between solidus *S* and liquidus *L* over the whole composition range. Additionally, in a few cases of ideal mixed systems, a partial section exists within the *T*–*x* projection near one of the components, like in the system CdTe–CdSe (Fig. 3.8b). In the region of mole fraction $0 < x_{CdSe} < 0.18$, liquidus and solidus are de facto coinciding. Principally, mixed crystal systems with small *L*–*S* lenses are proving to be advantageous at crystal growth from melt due to the minimal effect of segregation (the ratio k_0 in Fig. 3.6b) and, thus, reduced chemical composition inhomogeneity along the crystallization direction.

In the past, often, the ideal solution model has been applied to describe the molar Gibbs free energy of a material system. However, it has been recognized that the ideal solution model mostly suffers from a lack of accuracy, as already reported by *Hildebrand* in 1927 and by *Guggenheim* in 1932. They introduced the *regular solution model* based on the framework of statistical mechanics for binary systems, which are discussed in the next chapter.

3.2.3 Real mixed systems

Deviations from ideality are the usual case. The atoms or molecules of the substances to be mixed show enhanced energetic interaction (attraction or repulsion) either between

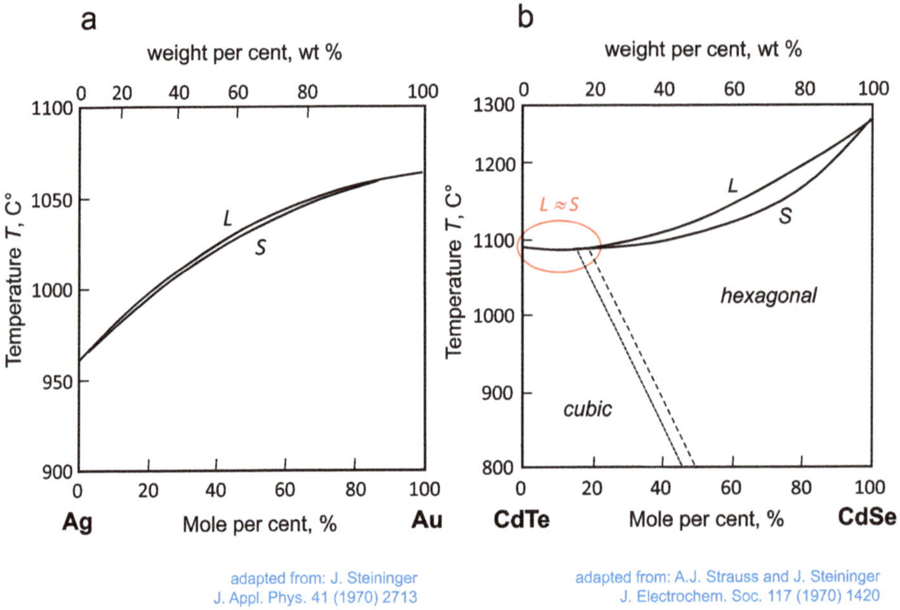

Fig. 3.8: Systems with nearly ideal mixing in the liquid and solid phases (*with permission of AIP Publishing (a) and The Electrochem. Soc.(b)*).

each other or among themselves. Now, in comparison to ideal mixed systems, a real mixed one shows not only a given molar entropy of mixing but also a changed enthalpy of mixing and, thus, in addition to $\Delta s_m \neq 0$, it is also $\Delta h_m \neq 0$. Therefore, the functional course of the molar free energy $g(x) = \mu(x)$ of a real mixed crystal, introduced by eq. (3.22), differs quantitatively from that of an ideal mixed crystal as

$$g(x) = \mu(x) = \sum_{i=A}^{Z} \mu_{i0} x_i + (\Delta h_m - T\Delta s_m) \tag{3.35}$$

Substituting the expressions for Δh_m and Δs_m, according to eqs. (3.24) and (3.25), becomes

$$\mu(x) = \sum_{i=A}^{Z} \mu_{i0} x_i + \sum_{i=A}^{K} x_i \, \Delta h_{im} - RT \sum_{i=A}^{K} x_i \ln x_i \tag{3.36}$$

A considerable simplification of eq. (3.36) takes place when we have to do it with a system *of regular solution*, which is quite usual for the case of dilute solutions, where the matrix component is markedly in excess. Especially, if the atomic (or molecular) sizes and electronic structures of the components (e.g., A and B) are similar, then the distribution will be nearly random, and the configurational entropy will be nearly ideal. Thus, the solution is formed by *random mixing of components* without strong specific interactions and its behavior differs from that of an ideal solution only modestly. Although the entropy of mixing does not have to be taken into account ($\Delta s_m \approx 0$; note, it is with some share that it is still integrated in activity), the system is further nonideal due to a non-zero enthalpy of mixing ($\Delta h_m \neq 0$), and eq. (3.36) becomes

$$\mu(x)^{\text{reg}} = \sum_{i=A}^{Z} \mu_{i0} x_i + \sum_{i=A}^{K} x_i \, \Delta h_{im} \tag{3.37}$$

where $\Delta h_{im} = \Delta h_{im}^{\text{reg}}$ is the partial molar enthalpy of a regular solution, the so-called thermal tone, generating or releasing during the mixing process. According to eq. (3.29) applies

$$\Delta h_{im}^{\text{reg}} = RT \ln a_i = RT \ln x_i \gamma_i \tag{3.38}$$

Then, the total chemical potential of a regular mixing system consisting of two components A (matrix) and B (solvent) in eq. (3.37) is

$$\mu(x)^{\text{reg}} = \left(\mu_A^0 x_A + \mu_B^0 x_B\right) + RT(\ln x_A + \ln x_B) + RT(\ln \gamma_A + \ln \gamma_B) \tag{3.39}$$

where the first two terms on the right represent the ideal potential (mechanical mixing) and the last term, the excess potential, as was introduced by eq. (3.19).

The knowledge of the variation with temperature and composition of the activity a_i of component i, or the activity coefficient of γ_i is of primary importance for the thermodynamic treatment of solutions, especially mixed melts and mixed crystals. This is required for the determination of the partial molar Gibbs energy of mixing of the components, which is required for the determination of the equilibrium state of any mixing process that involves component i in the solid solution.

The activity coefficient γ_i is a unitless thermodynamic function. Its value varies around unity. An activity of 1 means that the component is pure and does not exhibit solid solution, or it is ideally mixable with another component. If $\gamma_i > 1$, a nonideal solution takes place, where the component i exhibits a positive deviation from Raoult's law, then $a_i > x_i$, and in the evaluation of its chemical potential, component i "acts as if" the solution contains more of components i than the mole fraction suggests. Similarly, if $\gamma_k < 1$, a nonideal solution is presented, where the component i exhibits a negative deviation from Raoult's law so that $a_i < x_i$, the component "acts as if" there is less of it present than the composition suggests.

It should be noted here that there is an extensive number of literature, textbooks, and internet links explaining the experimental and theoretical determination of the activity coefficient in more detail. Generally, the variation of a_i or γ_i with temperature and composition must be determined experimentally. One of the experimental methods is the determination of the partial vapor pressure of the solvent and solute p_i. In accordance with Raoult's law, the relative partial pressure p_i/p_{i0} (p_{i0} – equilibrium vapor pressures of the pure components) is proportional to the product of mole fraction and the coefficient of fugacity (≡ coefficient of activity) $p_i/p_{i0} = \gamma_i\, x_i$, which becomes $\gamma_i = (p_i/p_{i0})1/x_{i0}$.

Frequently, however, few or no mixture data are at hand, and it is necessary to estimate activity coefficients from some suitable prediction method. Unfortunately, only few truly reliable prediction methods have been established. A relatively simple theoretical model of activity is to assume that the free energy of mixing versus mole fraction of a binary system has a minimum symmetry at 50:50 solution and increases symmetrically toward each pure component following a parable curve $y = mx^2$. Then, for such a *symmetric solution model*, the activity coefficient γ_i in a binary system is related to the excess enthalpy of mixing as

$$\Delta h^{ex}_{Am}(x) = RT \ln \gamma_A = \Omega(1 - x_A)^2 = \Omega x_B^2 \tag{3.40}$$

$$\Delta h^{ex}_{Bm}(x) = RT \ln \gamma_B = \Omega(1 - x_B)^2 = \Omega x_A^2 \tag{3.41}$$

Or, universally, for multicomponent solutions

$$\Delta h^{ex}_{im}(x) = RT \ln \gamma_i = \Omega(1 - x_i)^2 = \Omega x_j^2 \tag{3.42}$$

where Ω *is the interaction parameter*, comparable with *energy of interaction* ω (*Margules parameter*). In detail, each atom in a fluid and solid phase interacts energetically with the surrounding atoms of different elements, which can be derived for a binary solution by the *model of quasi-chemical equilibrium* in the intensified form as

$$\Omega = ZN_A \left[\omega_{AB} - \frac{1}{2}(\omega_{AA} + \omega_{BB})\right] \tag{3.43}$$

with Z the number of nearest neighbors, N_A the Avogadro number, and ω_{AB}, ω_{AA}, and ω_{BB} the interaction energies of the three possible neighbor pairs A–B, A–A, and B–B, respectively. Note that interaction energies are attractive or repulsive forces between atoms or molecules. We must not confuse them with bonds! Of course, they are correlating with bond energies as following: the system requires a low interaction energy when the bond energy between pairs is strong. On the other hand, a large amount of energy is needed to bring together pairs with low mutual bond energies. Incidentally, the exact determination of the total interaction energies in a given mixed system proves to be a certain challenge requiring theoretical ab initio calculations and tuned spectroscopic analysis.

To simplify, the value of Ω is often assumed to be constant over a range of physical conditions. Somewhat more complicated is to acknowledge that it varies with temperature or even with pressure. The detailed background of eq. (3.43) and further derivations for the interaction parameter are given in the *Spec box 3.6*.

In eq. (3.43), one can differentiate between the following two general cases demonstrated in Fig. 3.9:

a) strong mutual bond energies between the components A and B but weak bonds inside both the partners A–A and B–B, needing a lower interaction energy between A and B, so that eq. (3.43) accordingly becomes

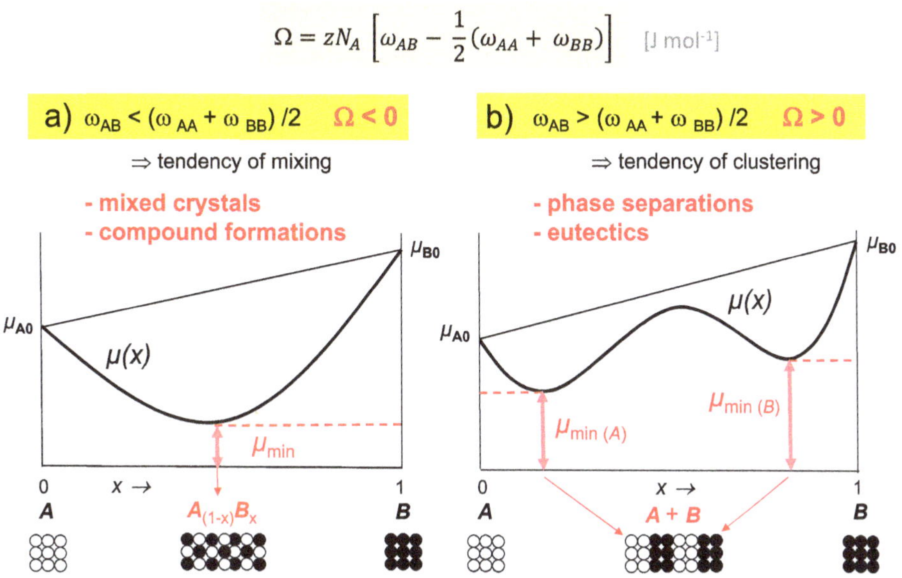

Fig. 3.9: Mixed solutions with a) attractive and b) repulsive forces between the components *A* and *B*.

$$\omega_{AB} < \frac{1}{2}(\omega_{AA} + \omega_{BB}) \quad \text{and} \quad \Omega < 0 \tag{3.44}$$

As a result, the system reduces the inner energy when each atom is surrounded by the other kind of atoms and it therefore tends to form a mixed fluid or solid (*mixed crystal*), sketched in Fig. 3.9a. In the particular case, the enhanced chemical reaction between the differing elements leads to the formation of *compounds AB*;

b) strong bonds inside each pure component *A–A* and *B–B* but only weak mutual bond energies between the differing components *A* and *B*, needing a higher interaction energy between *A* and *B* so that eq. (3.43) accordingly becomes

$$\omega_{AB} > \frac{1}{2}(\omega_{AA} + \omega_{BB}) \quad \text{and} \quad \Omega > 0 \tag{3.45}$$

As a result, the system tends to *phase separation* into both individual phases *A* and *B*, like in *eutectic* solid states or *monotectic* liquid states, for example (see Fig. 3.9b).

In comparison, an *asymmetric solution model* assumes that the excess enthalpy of mixing of a binary solution can be described by

$$\Delta h_{im}^{\text{ex}}(x) = x_A x_B (\Omega_{h1} x_B + \Omega_{h2} x_A) \tag{3.46}$$

That means, we have two interaction parameters instead of one in the symmetric model. Thus, a curve describing the free energy of mixing now has a distorted parabolic shape, minimizing at some composition other than $x = 0.5$ unless $\Omega_{h1} = \Omega_{h2}$. For this model, the activity coefficients are calculated as

$$\ln \gamma_A = x_B^2[\Omega_{h1} + 2x_A(\Omega_{h1} - \Omega_{h2})] \tag{3.47}$$

$$\ln \gamma_B = x_A^2[\Omega_{h1} + 2x_B(\Omega_{h1} - \Omega_{h2})] \tag{3.48}$$

In 1975, *Pelton and Thomson*, using a symmetric solution model, calculated for a hypothetical system *A–B* with regular solid and liquid phases, the topological change in the phase diagram via systematic changes of the interaction parameters Ω_S and Ω_L . The obtained Fig. 3.10 shows emphatically how the well-known principal types of *T–x* projections come into existence via such thermodynamic considerations – a really impressive picture for each crystal grower!

Already in 1948, *Redlich and Kister* extended the interaction energy term for binary and multicomponent systems to adapt the model still better to experiments by estimating the standard deviation between the experimental and calculated data by the so-called *sub-regular model*. This analysis can be used to fit activities in two-component mixtures over the entire concentration range, and to calculate the activities of both components. The variation of the parameters with the composition and temperature of the mixtures has been discussed in terms of the molecular interactions in these mixtures. The molar excess Gibbs free energy is expressed as

$$\frac{g^{ex}}{RT} = x_A x_B\left[B + C(x_A - x_B) + D(x_A - x_B)^2\right] \tag{3.49}$$

where *B*, *C*, and *D* are constants. In terms of the ratio of activity coefficients, the *Redlich–Kister equation* is

$$\ln\frac{\gamma_A}{\gamma_B} = B(x_B - x_A) + C(6x_A x_B - 1) + D(x_B - x_A)(1 - 8x_A x_B) \tag{3.50}$$

whereas the determination of the constants *B*, *C*, and *D* involves plotting of $\ln(\gamma_A/\gamma_B)$ as a function on x_A and reading the ordinate values from the curve at various preselected x_A values, for example, when $x_A = 0.5, C/2 = \ln(\gamma_A/\gamma_B)$.

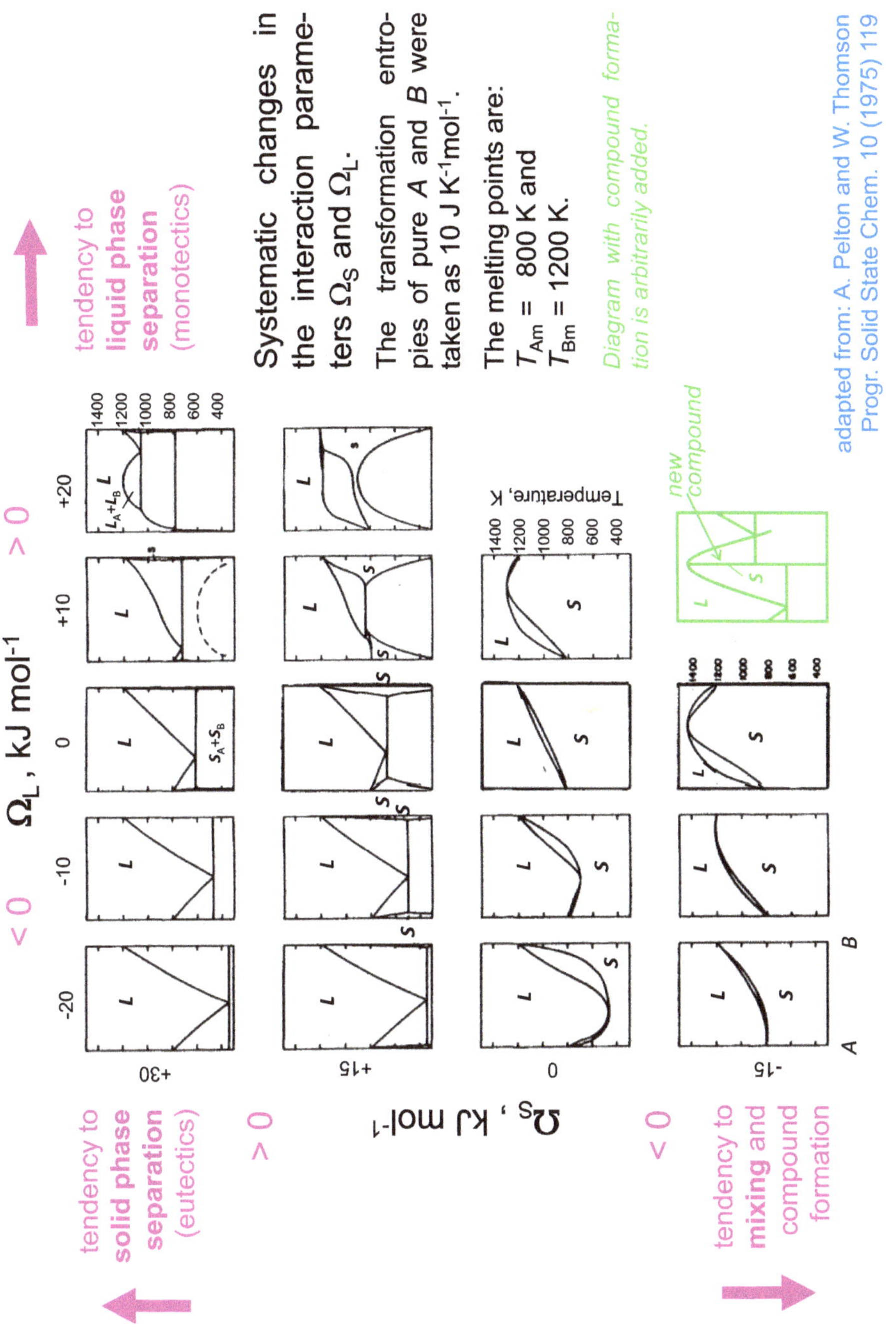

tendency to liquid phase separation (monotectics)
tendency to solid phase separation (eutectics)
tendency to mixing and compound formation
Systematic changes in the interaction parameters Ω_S and Ω_L.
The transformation entropies of pure A and B were taken as 10 J K-1mol-1.
The melting points are:
T_{Am} = 800 K and
T_{Bm} = 1200 K.
Diagram with compound formation is arbitrarily added.
new compound
adapted from: A. Pelton and W. Thomson Progr. Solid State Chem. 10 (1975) 119
Ω_L , kJ mol-1
Ω_S , kJ mol-1
Temperature, K

Fig. 3.10: Topological change in the *T–x* phase projections as function on the interaction parameter in regular solid and liquid phases of *A–B* systems (*with permission of Elsevier*).

Spec box 3.6: Predictions of the interaction parameter

Several approaches are discussed in the literature. Here, we show only some selected models being applicable, especially for crystallization processes.

i) Henrian ideal solution

In case, a solution is sufficiently diluted in one component, one can approximate

$$\mu_i \approx RT \ln \gamma_i \tag{B3.6-1}$$

by its value in an infinitely dilute solution. That is, if the mole fraction of the solute x_i is small, one can set $\gamma_i = \gamma_{i0}$, where γ_{i0} is the *Henrian activity coefficient* at $x_i \approx 0$. Thus, for sufficiently dilute solutions, one can assume that γ_i is independent of composition. Physically, this means that in a very dilute solution, there is negligible interaction among solute particles. Hence, each additional solute particle added to the solution makes the same contribution to its excess Gibbs energy so that $\mu_i = \partial G/\partial n_i = \text{const}$. According to the *Gibbs–Duhem equation* for a binary solution, $x_A\, d\mu_A{}^{eq} + x_B\, d\mu_B{}^{eq} = 0$, a negligible potential change of the solute (A) $d\mu_A{}^{eq} \approx 0$ equates to $d\mu_B{}^{eq} \approx 0$ of the solvent (B), leading to the ideal solvent behavior with constant $\gamma_B \approx 1$. Finally, for dilute solutions with $x_B \approx 1$, *Henry's law* is applied

$$\gamma_A = \gamma_A^0 = \text{const}; \quad \gamma_B \approx 1 \tag{B3.6-2}$$

Henrian activity coefficients can usually be expressed as functions of temperature

$$RT \ln \gamma_i^0 = a - bT \tag{B3.6-3}$$

where a and b are constants. If data are limited, it can further be assumed that $b \approx 0$ so that $RT \ln \gamma_{i0}$ = const.

ii) Model of quasi-chemical equilibrium (QCE)

This belongs to an *atomistic interpretation*. In pure components, each atom possesses Z nearest neighbors, i.e., Z $(A–A)$ or Z $(B–B)$ bonds with the interchange energy per atom $Z/2\omega_{AA}$ or $Z/2\omega_{BB}$, respectively. Multiplying by the Avogadro's constant, N_A becomes the interchange energies per mol, $N_A\,(Z/2)\,\omega_{AA}$ or $N_A\,(Z/2)\,\omega_{BB}$. The probability of $A–A$ or $B–B$ bonds within two neighboring sites is $(1–x_B)^2$ or $x_B{}^2$, respectively. Thus, the quantity of $A–A$ and $B–B$ bonds is $N_A\,(Z/2)\,(1–x_B)^2$ and $N_A\,(Z/2)\,x_B{}^2$.

Besides homogeneous bonds $A–A$ and $B–B$, similar to a chemical reaction, in a mixed system also, heterogeneous $A–B$ bonds are formed. According to the quasi-chemical treatment, the statistical quantity of $A–B$ bonds per mol is $N_A\, Z\, x_B x_A = N_A\, Z\, x_B(1–x_B)$, with x_A, x_B the mole fractions.

Now, the total inner energy of a mixed crystal is assumed to be the sum of the interchange energies of the three possible types of neighborly bonds, $A–A$, $B–B$, and $A–B$, given as

$$\omega_{A-B}^{\Sigma} = \omega_{AA} + \omega_{BB} + \omega_{AB} \tag{B3.6-4}$$

Note, in this approach, the influence of heat oscillations on the atomic arrangement and the strain energy between atoms are neglected and it is treated only spatially as *configurational energy*. Inserting the three bond quantities from the above into eq. (B3.6-4), the total interchange energy is

$$\omega_{A-B}^{\Sigma} = \frac{1}{2} N_A Z \omega_{AA}(1-x_B)^2 + \frac{1}{2} N_A Z \omega_{BB} x_B^2 + N_A Z \omega_{AB} x_B(1-x_B) \tag{B3.6-5}$$

and after adjusting this formula, one obtains

$$\omega_{A-B}^{\Sigma} = \frac{1}{2} N_A Z [\omega_{AA}(1-x_B) + \omega_{BB} x_B] + N_A Z \left[x_B(1-x_B) \left(\omega_{AB} - \frac{\omega_{AA} - \omega_{BB}}{2} \right) \right] \tag{B3.6-6}$$

whereas the second term on the right describes the *mixing interchange energy*, $N_A\, x_B\, x_A\, \Omega$, with the interaction parameter Ω as has been already ad hoc presented in eq. (3.43).

$$\Omega = ZN_A\left[\omega_{AB} - \frac{1}{2}(\omega_{AA} + \omega_{BB})\right] \tag{B3.6-7}$$

Finally, using eq. (3.42), the activity coefficients of the constituents are calculable by

$$RT \ln \gamma_i = \Omega(1 - x_i)^2 = \Omega x_j^2 \tag{B3.6-8}$$

Solutions with constant interaction parameter Ω are classified as *strictly regular*. Taking into account the temperature dependence, a *quasi-regular solution model* is

$$\Omega = a - bT \tag{B3.6-9}$$

comparable with (B3.6-3), whereupon it can be inserted $a = RT \ln \gamma_i / x_j^2$ and $b = \Delta s_m / x_j^2$. If $b = 0$, a strictly regular case takes place. An athermal solution occurs if $\Omega = -bT$.

iii) Model of delta-lattice parameter (DLP)
In 1972, *Stringfellow* developed a semiempirical expression for the calculation of the interaction parameter, based on the difference in the lattice constants of the constituents in *binary* $A_{1-x}B_x$ and *pseudo-binary ternary* semiconductor mixed crystals $AB_{1-x}C_x$. In this model, the enthalpy of mixing is related to the effect of composition on the total energy of the bonding electrons. This leads to an approximate representation of Ω in terms of only the lattice parameters of the two binary constituents *AB* and *AC*:

$$\Omega = 4.375K(\Delta a)^2 / \bar{a}^{4.5} \tag{B3.6-10}$$

with Δa the lattice parameter difference, $\bar{a}$ the average lattice parameter, and *K* an adjustable constant that can be obtained by fitting to the experimental data according to $\Delta H_{at} = K\, a^{-2.5} > 0$ (ΔH_{at} – enthalpy of atomization, i.e., the amount of enthalpy change when a compound's bonds are broken and the component elements are reduced to individual atoms). *K* proved to be the same for all III–V alloys, yielding a value of 1.15×10^7 cal/mole $Å^{-2.5}$. The Δa^2 dependence in eq. (B3.6-10) suggests that the main effect in the DLP model is the strain energy associated with deformation of the bonds in the alloy. The later developed *valence force field (VFF) mode* specifically considers the short-range energy required to stretch and bend the bonds without the adjustable parameter.

iv) Model for ionic solutions
For the solution of substances that ionized in solution, the activity coefficients of the cation and anion cannot be experimentally determined independently of each other because solution properties depend on both ions. However, single-ion activity coefficients can be calculated theoretically. Once, one knows the molar concentration of the free ion c_i (~ mole fraction x_i), it is converted to the activity a_i by the *free-ion activity coefficient* γ_i

$$a_i = \gamma_i\, c_i \tag{B3.6-9}$$

whereas γ_i corrects for electrostatic shielding by other ions and, hence, γ_i depends on the *ionic strength* I (i.e., the concentration of electrical charge), which is

$$I = \frac{1}{2}\sum_i z_i^2\, c_i \tag{B3.6-10}$$

Here, z_i is the charge of ion *i*. The sum is taken over all ions in the solution. Due to the square of z_i, multivalent ions contribute strongly to the ionic strength.

A simple expression for the activity coefficient in electrolyte solutions is given by the *model of Debye–Hückel (1923)* by taking into account long-range electrostatic interactions between ions

$$\lg \gamma_i = -A z_i^2 \sqrt{I} \tag{B3.6-11}$$

where A is the parameter depending on aqueous solution on temperature T, and the dielectric constant ε as

$$A = 1.82 \times 10^6 (\varepsilon T)^{-3/2} \tag{B3.6-12}$$

This model provides good results only at very dilute concentrations, so a variety of improvements have been made for more concentrated systems by *Pitzer (1991)*.

3.2.4 Selected binary phase diagrams

3.2.4.1 Mixed crystals with nearly ideal solid solution

Mixed bulk crystals and epitaxial layers are of highest importance, such as the semiconductor alloys $Ge_{1-x}Si_x$, $In_{1-x}Ga_xAs$, $Ga_{1-x}Al_xN$, and $Cd_{1-x}Zn_xTe$, for example. By changing the composition the lattice parameter and electrical qualities (width of band gap) can be adjusted. Usually, a high compositional homogeneity (x – uniformity) is required, especially, when mixed crystalline substrates or detectors are used. On the other hand mixed crystals with linear variation of the "lattice constant" (called delta crystals) have enhanced mechanical stability (Ginzburg–Landau theory) and are applicable as lenses in X-ray optics. The difficulty of this concept is achieving linear variation of the lattice constant with high accuracy and without loss of single crystallinity. In both areas, mixed crystals with nearly ideal solution behavior are favored.

An ideal solution A–B means a system with complete miscibility in both the liquid and solid phase within the whole composition range, $0 < x < 1$. In the low pressure T–x projection of such a phase diagram, there are the three fields, liquid, liquid + solid, and solid, separated by two boundaries known as the *liquidus L* and *solidus S* (Fig. 3.6). Using the above derived thermodynamic knowledge, the S and L courses of such ideal systems can be calculated analytically. With respect to the component $i = B$ added to the matrix component A, the equilibrium between solid (S) and liquid (L) phases is given by the equality of the chemical potential of each phase from eq. (3.29)

$$\mu_{BS}(x_{BS}, T) = \mu_{BL}(x_{BL}, T) \tag{3.51}$$

$$\mu_{BS}^0 + RT \ln x_{BS}\gamma_{BS} = \mu_{BL}^0 + RT \ln x_{BL}\gamma_{BL} \tag{3.52}$$

which, after adjusting, becomes

$$RT \ln\left(\frac{x_{BS}\gamma_{BS}}{x_{BL}\gamma_{BL}}\right) = \mu_{BL}^0 - \mu_{BS}^0 = \Delta\mu_B^0 = \Delta h_B^0 - T\Delta s_B^0 = 0 \tag{3.53}$$

where h_{B0} and $s_{B0} = h_{B0}/T_{Bm}$ are the intensive standard enthalpy and entropy of the added pure component B with melting temperature T_{Bm}, respectively. Hence, eq. (3.53) is then

$$\ln\left(\frac{x_{BS}\gamma_{BS}}{x_{BL}\gamma_{BL}}\right) = \frac{\Delta h_B^0}{R}\left(\frac{1}{T} - \frac{1}{T_{Bm}}\right) \tag{3.54}$$

Since the solutions in both phases are ideal, the activities $\gamma_{BS,L}$ are ≈1 and eq. (3.54) is transformed into the *van Laar equation*:

$$\frac{x_{BS}}{x_{BL}} = k_0 = \exp\left[\frac{\Delta h_B^0}{R}\left(\frac{1}{T} - \frac{1}{T_{Bm}}\right)\right] \tag{3.55}$$

where k_0 is the (thermodynamic) ***equilibrium distribution (segregation) coefficient*** of added component B in matrix A. Note that in the case of mixed crystal. Ge(A)–Si(B) is $k_{0\mathrm{Si}} \geq 1$ for all values of the mole fraction x_{Si}, as demonstrated by the insertion in Fig. 3.11.

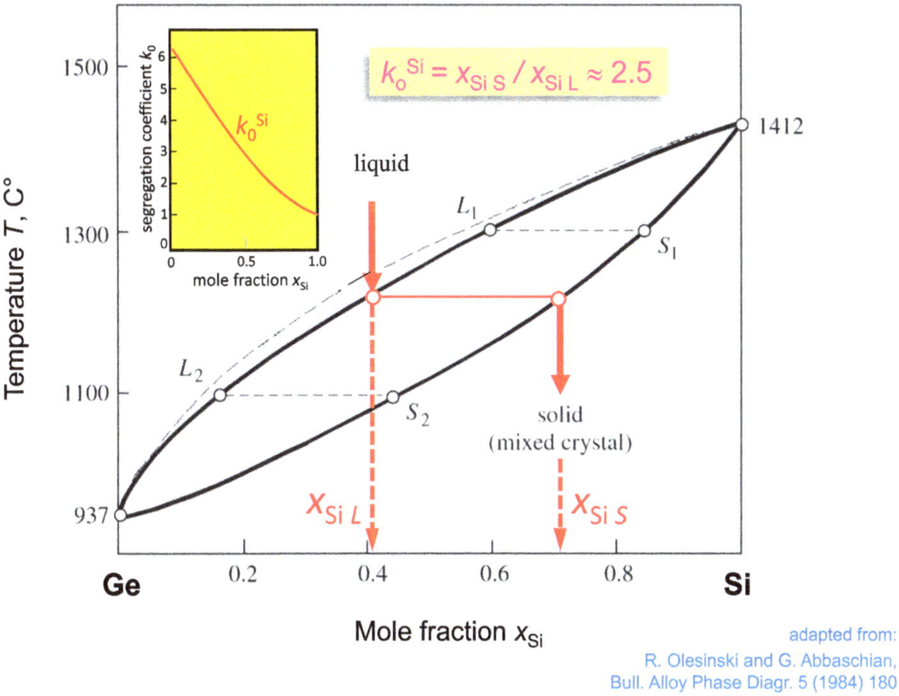

Fig. 3.11: *T–x* projection and $k_0(x)$ course of the near ideal system Ge–Si. Full lines were calculated, dashed line is experimental (*with permission of Springer Nature*).

Using $x_{A\ S,L} + x_{B\ S,L} = 1$, the equation of the *liquidus* is

$$x_{AL}\exp\left[\frac{\Delta h_A^0}{R}\left(\frac{1}{T} - \frac{1}{T_{Am}}\right)\right] + x_{BL}\exp\left[\frac{\Delta h_B^0}{R}\left(\frac{1}{T} - \frac{1}{T_{Bm}}\right)\right] = 1 \tag{3.56}$$

and, the equation of the *solidus* is

$$x_{AS}\exp\left[-\frac{\Delta h_A^0}{R}\left(\frac{1}{T}-\frac{1}{T_{Am}}\right)\right]+x_{BS}\exp\left[-\frac{\Delta h_B^0}{R}\left(\frac{1}{T}-\frac{1}{T_{Bm}}\right)\right]=1 \tag{3.57}$$

For instance, let us calculate the T–x projection of the system Ge–Si by applying eqs. (3.56) and (3.57). The near-linear variation of the lattice constant with composition (*Vegard's rule*) leads us to expect a good solubility between both material partners. After replacing $x_{A\ L,S}$ by $(1–x_{B\ L,S})$ and inserting $\Delta h°_{Ge} = 32$ kJ/mol, $\Delta h°_{Si} = 50$ kJ/mol, $T_{mGe} =$ 1210 K, and $T_{mSi} = 1685$ K, quite a good agreement with the experimental data is obtained (Fig. 3.11). However, as can be seen, the experimental liquidus curve is somewhat above the calculated one, indicating that the solutions are not quite ideal. Indeed, several precise treatments within the literature provided the evidence that the liquidus and solidus data could be best approximated when the activity coefficients $\gamma_{B\ L,S}$ in eq. (3.54), instead of unity, are expressed by the interaction parameters $\Omega_L = 1.615$ kcal/g-atom and $\Omega_S = 1.210$ kcal/g-atom, as in eqs. (3.47) and (3.48).

The most important challenge for the crystal grower is to manage mixed crystals A_xB_{1-x} or $A_xB_{1-x}C$ etc. with homogeneous composition throughout their whole length and diameter. The difficulty lies in the high mutual solubility of the constituting components over the whole mole fraction causing a permanent *effect of segregation* during the growth process, due to the marked difference between the liquidus and solidus temperatures in their phase diagrams (see Fig. 3.6). As a result, throughout the directional crystallization, both the solvent and solute are gradually enriched in the remaining melt and increasing crystal parts. Additionally, the appearance of a massive *diffusion boundary layer* at the propagating solid–liquid interface always entails the danger of constitutional supercooling (lecture parts III and IV). In this context, the exact knowledge of the segregation coefficient, introduced for an ideal mixed system by eq. (3.55), is of particular importance as will be discussed in Section 3.2.4.4.

3.2.4.2 Systems with congruent melting compounds

3.2.4.2.1 Stoichiometry and region of homogeneity

As mentioned in Section 3.2.3 (eq. (3.43)), strong mutual bond energies between the components A and B but weak bonds inside the both partners A–A and B–B result in lower interaction energy between A and B so that the interaction parameter becomes $\Omega < 0$. In other words, the material system reduces the inner energy when each atom surrounds the other kind of atoms. One realization possibility is the ideally mixed system with randomly (*disordered*) arranged atomic structure as we presented in Section 3.2.4.1. However, with increasing attraction between the unlike atoms, a disorder–order transition with formation of an intermediate phase (*compound AB*) can take place where the atoms occupy well-defined (*ordered*) lattice sites. Figure 3.12a shows such a situation. Mostly, intermediate phases are characterized by enhanced covalent and ionic bond percentages. Therefore, the free energy, including entropy and enthalpy, of such phases are typically low, causing their high stability. Intermediate phases apply to many important dielectric (e.g., alkali halides, corundum, and garnets) and semiconductor compounds (e.g., III–V, II–VI, and IV–VI).

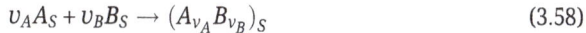

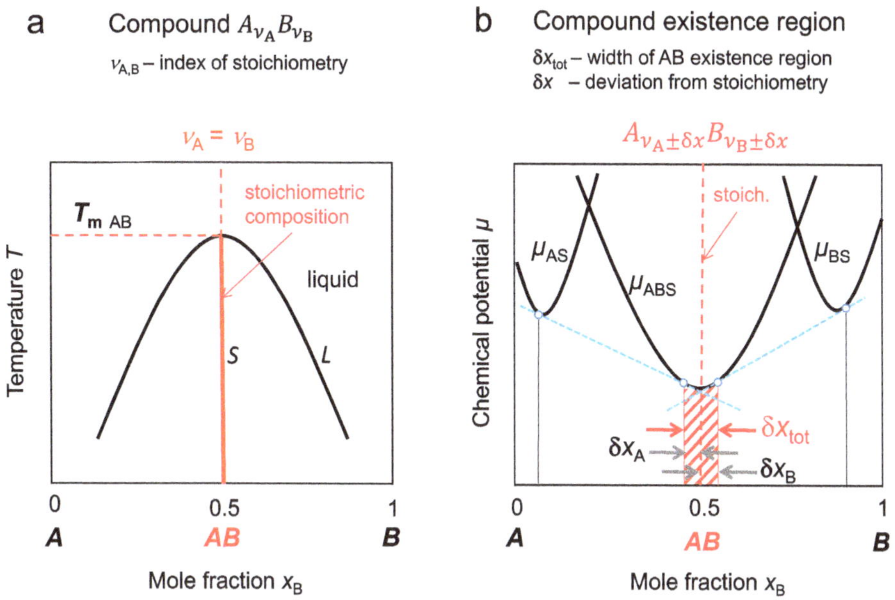

Fig. 3.12: Sketches to illustrate a compound stoichiometry *AB* (a) and the formation of the width of an existence region (b) by the tangent constructions between chemical potential minima of pure components *A* and *B* and an related compound *AB* at a given temperature.

A completely ordered compound structure exists only if the ratio of the number of atoms *A* and *B* is equal to the ratio of relatively small integers υ_A and υ_B. The compound can then designated by the formula $A_{\upsilon A}B_{\upsilon B}$. Considering a solid phase consisting of *A* and *B* partners, the following chemical reaction for the same integers $\upsilon_A = \upsilon_B$ is

$$\upsilon_A A_S + \upsilon_B B_S \longrightarrow (A_{\nu_A} B_{\nu_B})_S \qquad (3.58)$$

Such examples within the class of semiconductors are GaAs, InP, GaN, ZnSe, and CdTe. Equally proportioned dielectric compounds are NaCl, CsCl, and LiF. Further, there are many compounds of unequal proportions showing different integers, such as $\upsilon_A = 2\upsilon_B$ (e.g., Mg_2Si) or $\upsilon_A = 2/3\upsilon_B$ (e.g., Bi_2Te_3 and Al_2O_3). Also, multiple compounds consisting of more than two basic elements are composed of unequal integers like $LiNbO_3$ or $La_3Ga_5SiO_{14}$, for example.

All these compounds have one thing in common: they are of *stoichiometric composition* (see Fig. 3.12a). Stoichiometry expresses the law of conservation of mass where the total mass of the reactants equals the total mass of the products. Thus, stoichiometry denotes a strong chemically defined singular composition. Unfortunately, in the crystal growth literature, very often an incorrect description is used, whereupon a given compound of "different or various stoichiometry" can be crystallized. Of course, correctly it should be "different deviations from stoichiometry" (see below).

Figure 3.12b shows the scheme of the related $\mu(x)$ curves for all stable solid phases, A_S, AB_S, and B_S in a binary system, with a formed compound at a given temperature. Note, the minimum of the chemical potential equated to the free molar energy of the compound $\mu_{AB} = g_{AB}$ is not pointed but shows a comparably broad (dashed) area. This is due to some sideward contacts of the two equilibrium tangents with the compound potential curve $\mu_{AB}(x)$. This means that the compound AB exists over a certain x region, marked in Fig. 3.12b, by the total width δx_{tot} and referred to as the *region of existence* for each given temperature, often also named *homogeneity region*. As a result, the boundaries of the homogeneity region are deviated from stoichiometry by the distance δx_A and δx_B, respectively. Such a situation is a characteristic feature of most compounds showing certain solid solubility of both elementary components. From the figure, it follows that δx_{tot} would decrease with the tapering of the μ_{AB} (x) curve, expressing an increasing attraction of the opposite compounds or rather the bond energy between them. The appearance of a homogeneity region can be explained thermodynamically through the principle of Gibbs free energy minimization [see Section 2.1; eq. (2.2)], whereupon the entropy part S strives to increase G by inducing a certain degree of disorder. This is realized by introducing some imperfections in the form of *intrinsic (native) point defects*, established by addition of excess A or B atoms into the AB compound lattice. For that, three implementation scenarios are available: (i) *interstitial* positioning of the excess atoms, (ii) *vacancies* within the sublattice of the shortage component, and (iii) exchange of a lattice place by an opposite atom, designated as *antisite* (more details will be given in the lecture part IV). In result, the compound AB proves to be a *partly disordered solid phase*, existing not only at stoichiometric composition but within a region of existence, bordered by solidus lines, i.e., the courses of solubility limits for A or B in AB at given temperature and pressure.

Figure 3.13 presents the scheme of a T–x projection of a binary eutectic system with a AB compound that has a magnified existence region δx_{tot}, with deviations of the solidus curves from stoichiometry δx_A and δx_B. The μ–x projections for a liquid and a solid state at given temperatures are added left. Both solidus and liquidus are hyperbolae with asymptotes intersecting at one point on the compound axis. Due to the consistency of both liquid and solid composition, this singularity is denoted as *congruent melting point* T_{cmp} (CMP). Considering a congruent crystallization process, the solid compound is formed by the reaction of two liquid components, and eq. (3.58) becomes

$$\upsilon_A A_L + \upsilon_B B_L \rightarrow (A_{\nu_A} B_{\nu_B})_S \tag{3.59}$$

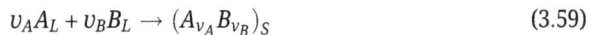

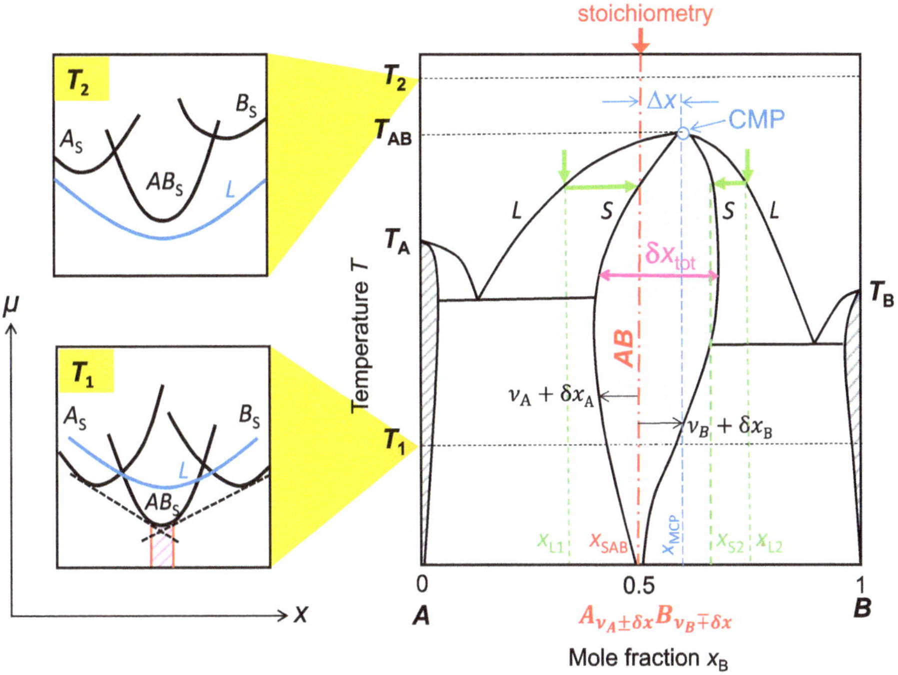

Fig. 3.13: Sketches of a widened existence region with congruent melting point (CMP) of a virtual compound *AB* within a binary eutectic system and two *μ*–*x* projections for the liquid and solid state, to illustrate the characteristic deviation of CMP from stoichiometry and the retrograde course of the boundary (solidus) of the existence region. Green arrows show the crystallization path on both sides of the existence region. Stoichiometry is only met by the left path (L - liquidus, S - solidus).

The *T*–*x* figure of the homogeneity region need not be symmetrical, and as a consequence, the maximum at which solid and liquid are identical does not necessarily meet at the stoichiometric composition. This is due to the differences in the energies of formation of the point imperfections. As a result, the minimum of the chemical potential (free molar energy) of the compound μ_{AB} is shifted to the side of that imperfection, the formation of which requires the least energy. In Fig. 3.13, this is assumed for the *B*-rich side. Hence, the degree of deviation of CMP from stoichiometry Δx corresponds to the intrinsic disorder. For instance, in GaAs, the CMP is deviated to the As-rich side at $\Delta x_{As} \approx 8 \times 10^{-3}$ at % (10^{18} excess As atoms per cm^3) and in CdTe, the CMP is located at Te excess of $\Delta x_{Te} \approx 1 \times 10^{-3}$ at % (1.5×10^{17} cm^{-3}). Consequently, the experimental achievement of a stoichiometric compound crystal by crystallization is always associated with the phenomenon of segregation due to the compositional difference between solidus and liquidus, outside the CMP. Two possible practical cases are inserted in Fig. 3.13 by green arrows. When the crystallization starts right of the CMP, e.g., at liquid composition x_{L2}, the resulting solid composition is always located outside the stoichiometry, like here at x_{S2}. As a result, the *AB* compound would contain an excess of *B*. On the other hand, to achieve a stoichiometric *AB* compound of x_{SAB}, the crystallization should start left of CMP at x_{L1}. Clearly however, without thermodynamic in situ stabilization, this starting composition is not kept constant during an unidirectional solidification process but would drift more and more toward the excess of *A* by shifting down along the liquidus course.

The extension of the homogeneity region δx_{tot} depends on the interaction energy between the *A* and *B* atoms as well as on the formation enthalpies of nonstoichiometry-related point defects. All this, together, leads to very differing and mostly asymmetric shapes of the homogeneity region. Due to the decreasing solubility with decreasing temperature, down to 300 K, the solidus curves of the homogeneity region are retrograde. This fact is of great importance for the *nucleation of precipitates* of the excess component *A* or *B* during the cooling down process of an as-grown compound crystal (ch. 5.2.4).

Considering the degree of deviation from stoichiometry at a given temperature $\pm\,\delta x$ (see Figs. 3.12b and 3.13), the correct expression of eq. (3.59) is then

$$(\upsilon_A \pm \delta x_A)A_L + (\upsilon_B \mp \delta x_B)B_L \longrightarrow (A_{\nu_A \pm \delta x_A} B_{\nu_B \mp \delta x_B})_S \tag{3.60}$$

Note, a compound with formula $A_{\nu_A} B_{\nu_B + \delta x_B}$ would refer to the deviation of the whole existence region from stoichiometric composition toward *B* excess as is really the case in SnTe and GaAs, for example (see Fig. 3.14a and c).

Figure 3.14 summarizes some forms, extents, and numbers of existence regions in selected binary and quasi-binary systems. Semiconductor compounds (a–c) are characterized by relative small homogeneity regions. In contrast, there are oxides and intermetallic systems with large extensions on either side of the stoichiometry, like $LiNbO_3$ (d) or AlNi (e). Even the intermetallic system Al–Ni is an instructive example of the formation of numerous subcompounds, in addition to AlN such as Al_3Ni_2, Ni_5Al_3, and Ni_3Al with incongruent melting points (see Section 3.2.4.3) and relatively wide existence regions. An example of a complex intermediate compound is the quasi-binary system Cu_2S–In_2S_3 (f). The homogeneity region of $CuInS_2$ is subdivided into three solid phases ξ, δ, and γ, to be passed through during cooling down – a real challenge for the crystal growth of this interesting photovoltaic material.

As was already mentioned above, the shape of the homogeneity region is the result of the energetic superposition of all intrinsic point defects at a given temperature. Of course, the addition of foreign atoms as dopants or impurities of an adequate concentration does slightly affect the configuration of the existence region too. Since at high temperatures, near to the melting point, intrinsic and extrinsic point defects are isolated and usually electrically charged, the width of the existence region corresponds to the principle of electrical neutrality of the crystal.

Table 3.1 compiles some maxima of phase extents, i.e., the total width δx_{tot} of selected compounds (compare with Figs. 3.12–3.14). The higher the defect formation energies, the smaller are δx_{tot} typically for semiconductor compounds. For physical and practical requirements, the mutual conversion of the mole fractions (moles per cent) into the concentration of atoms (molecules) [*C*] per cm^3 is feasible by the equation

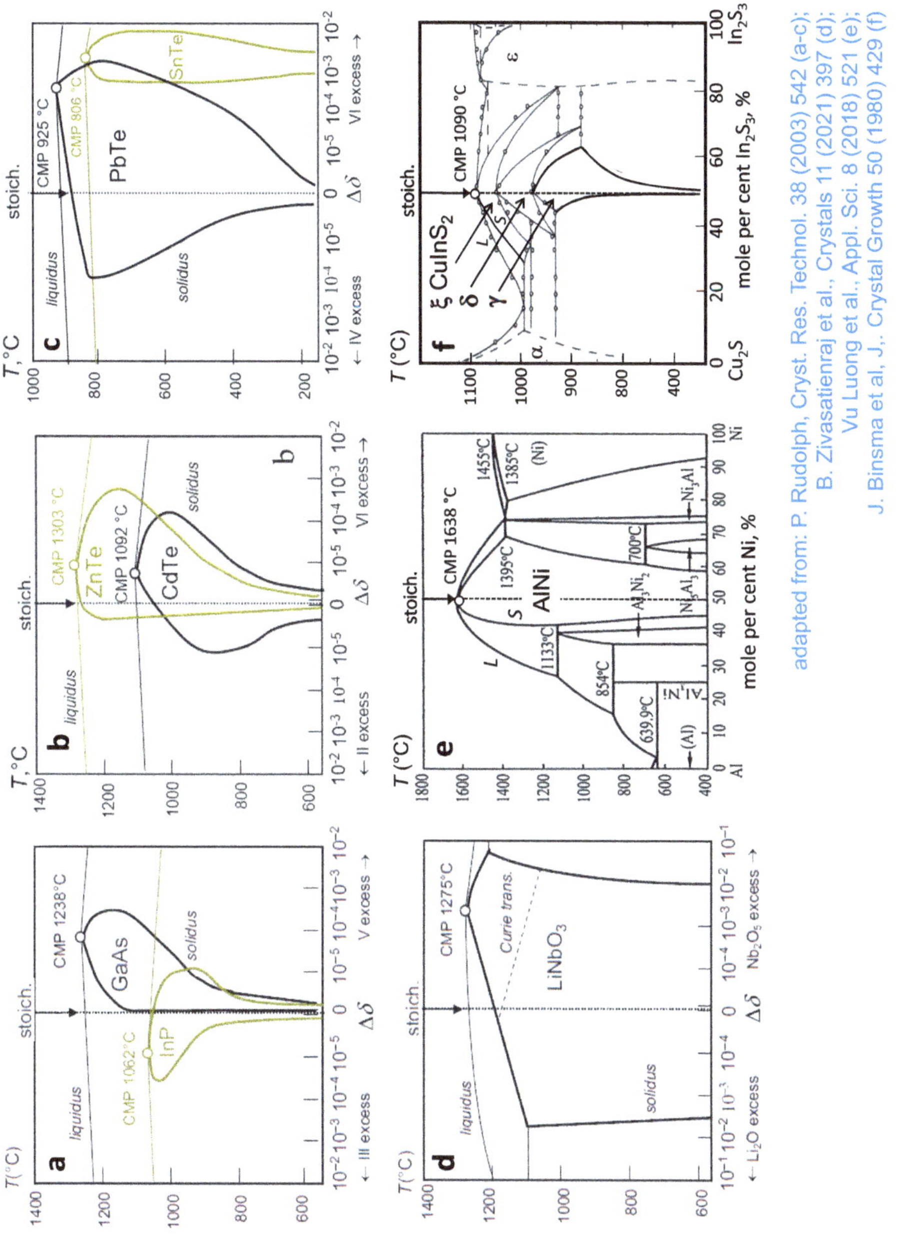

Fig. 3.14: Selected real existence regions of quasi-binary systems with solely (a–d), diverse (e) and solid multiphase existence regions (f) (*with permission of Wiley and Sons (a–c), open accesses by MDPI (d,e), and permission of Elsevier (f)*).

$$[C] = \frac{\delta x_{tot}(\text{or } \delta x) N_A \rho}{M} \left[cm^{-3}\right] \tag{3.61}$$

(N_A – Avogadro's constant in per mol, ρ – density of the compound at given temperature in g/cm^3, and M – molecular mass of the compound in g/mol). Both total extension of the existence region δx_{tot} and deviations from stoichiometry δx are given in dimensionless mole fraction.

Tab. 3.1: Maximum widths of existence regions in selected material (*with permission of Springer Nature; supplemented values are public domains of Wikipedia*).

Compound	GaAs*	CdTe	PbTe	MnSb*	$LiNbO_3$	$MgAl_2O_3$*	AlNi	$CuInS_2$
max. δx_{tot}	2×10^{-4}	3×10^{-4}	2×10^{-3}	5×10^{-2}	3×10^{-2}	3×10^{-1}	8×10^{-1}	2×10^{-1}
T at δx_{tot} (°C)	1,150	1,000	800	700	1,100	2,000	1,400	875
ρ (g/cm^3)	5.2	6.2	7.8	4.9	4.6	3.6	4.3	4.7
M (g/mol)	144.64	240.01	334.8	176.70	147.85	126.27	85.67	242.49
$[C]_{max}$ (cm^{-3})	4×10^{18}	5×10^{18}	3×10^{19}	8×10^{20}	6×10^{20}	5×10^{21}	2×10^{22}	2×10^{22}

*The whole existence region is completely deviated from stoichiometry toward *A*-side (MnSb) or *B*-side (GaAs and $MgAl_2O_3$)

3.2.4.2.2 The problem of incongruent evaporation

Above the surface of an *ideal mixed melt A–B*, the total vapor pressure p_{AB} is equal to the sum of the partial vapor pressures of the two constituents A and B, given by *Raoult's* relations $p_A = p_A{}^\circ x_A$ and $p_B = p_B{}^\circ x_B$ if no other gases are present ($p_A{}^\circ$ and $p_B{}^\circ$ are the equilibrium vapor pressures of the pure components). Once the vapor phase consists of both components in an ideal solution, the total vapor pressure can be determined by combining Raoult's with *Dalton's law* of partial pressures to give (Fig. 3.15a)

$$p_{AB} = p_A^0 x_A + p_B^0 x_B = p_A^0 x_A + p_B^0 (1 - x_A) \tag{3.62}$$

where the summands represent the partial vapor pressures $p_A = p_A^0 x_A$ and $p_B = p_B^0 x_B$.

Many pairs of liquids and vapors show no uniformity of attractive forces. They form nonideal but *real mixings*. Raoult's law can still be adapted by incorporating the coefficients of fugacity ϕ_i and activity γ_i for the vapor and liquid, respectively (see also Section 3.2.1). Then, the real partial pressure of a given element i is $p_i = p_i{}^\circ \phi_i\, x_i$. Figure 3.15b and c shows two real mixed material systems, with deviations of the pressure curves from ideal cases being exemplary for many multicompound materials that are processed in crystal growth. Negative deviations from Raoult's law arise when the forces between the constituents in the mixture are stronger than the forces between pure components (Fig. 3.15b). On the other hand, a positive deviation occurs when the attraction between similar components is greater than between the dissimilar ones (Fig. 3.15c). Then, the

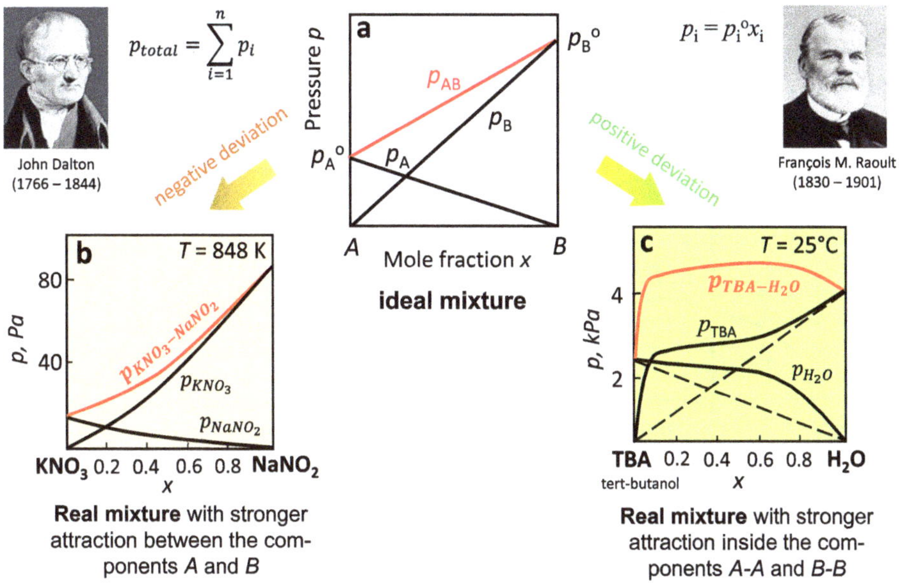

adapted from: V. Glazov et al., Rus. J. Appl. Chem. 75 (2002) 888 (b);
Y. Koga, Solution Thermodyn. and its Appl. to Aquous Sol. 2nd ed. (Elsevier 2017) p. 27 (c)

Fig. 3.15: Partial and total pressure curves in ideal (a) and real mixed systems with stronger attraction between the differing components (b) and inside the identical components (c). (*the portraits of J. Dalton and F.M. Raoult are public domains; with permission of Springer Nature (b) and Elsevier (c).*)

vapor over the mixed liquid is enriched by the most volatile component – a very often case at crystal growth of compounds.

It is usually the case that the p_i° values are quite different and, thus, the total pressure is essentially dependent on the partial pressure of the most volatile component quasi over the whole composition x_i. This is the case in the III–V systems. For instance, the partial pressures of arsenic p_{As_2}, p_{As_4} and phosphorous p_{P_2}, p_{P_4} along the liquidus are determined over nearly the whole Ga–As and In–P melt compositions. For instance, in the system Ga–As, the intersections with the partial gallium pressure p_{Ga} are very closed to the pure gallium (Fig. 3.16). Therefore, an uncovered Ga–As melt within a crucible standing in vacuum would dissociate by intense evaporation of the arsenic part, leaving behind nearly elementary liquid gallium. Thus, the Czochralski growth of GaAs (and other related compounds consisting of both high- and nonvolatile components) only became possible when the melt surface is covered by a liquid encapsulant (e.g., B_2O_3), as was successfully shown for the first time by *Mullin* et al. in 1965. Additionally, it is advisable for the growth vessel to fill by a pressurizing inert gas.

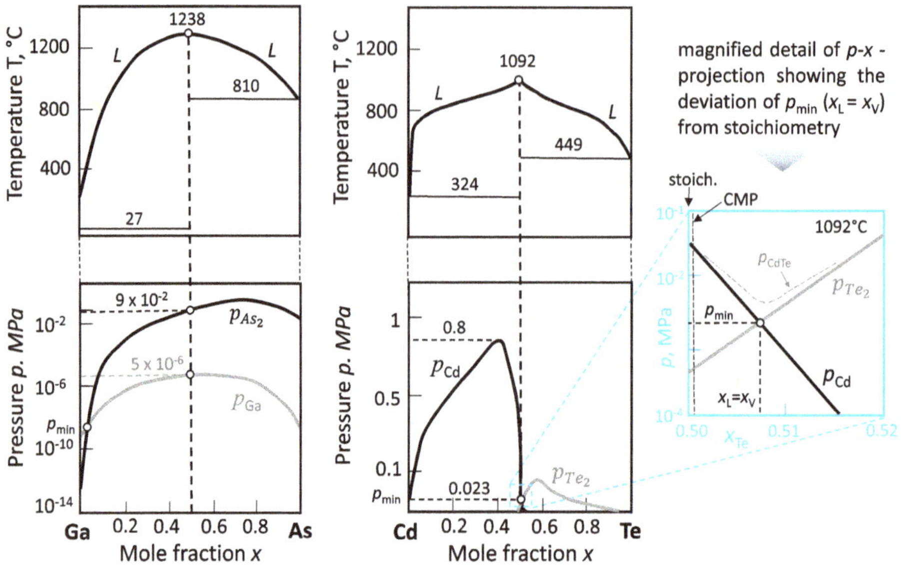

data from: H. Wenzl et al.,J. Crystal Growth109 (1991) 191; R. Brebrick, Comp. Phase Diagr. Thermochem. 34 (2010) 434; N. Yellin, S. Szapiro, J. Crystal Growth 73 (1985) 77; J. Greeberg, J. Crystal Growth 161 (1996) 1

Fig. 3.16: *p–x* projections of the partial pressures over the liquidus *L* in the *T–x* projection of the systems Ga–As and Cd–Te, showing the deviation of minimum total pressure from stoichiometry (*with permission of Elsevier*).

In comparison, there are numerous systems in which all consisting components show high-volatile partial pressures, like II–VIs. In Fig. 3.16, the *p–x* – phase projection with the partial pressure curves of p_{Cd} and p_{Te_2} along the liquidus of Cd–Te are presented. While p_{Cd} dominates on the Cd-rich side, p_{Te_2} controls the Te-rich side. The intersection point of both curves determines the composition of *congruent evaporation* $x_V = x_L$ which is distinct-deviated from stoichiometry in the direction of Te excess (at $x_{Te} \approx$ 50.7 at % $\approx 10^{20}$ excess Te atoms per cm^3). This crossing point coincides with the minimum of the total pressure curve and is designated as p_{min} (see the magnified detail graph in Fig. 3.16, right). It is noteworthy that at this point, the Cd–Te melt evaporates in its own composition without decomposing. Unfortunately, the $x_L = x_V$ equilibrium does not coincide with stoichiometry and lies significantly outside the CMP. One of the rare compounds with near equality between both CMP and p_{min} ($x_S = x_L \approx x_V = x_L$) proves to be SnTe. However, its whole region of homogeneity is markedly deviated from stoichiometry in the Te direction (Fig. 3.14c).

Table 3.2 compiles the deviations of the congruent melting point and p_{min} from stoichiometric composition of some selected semiconductor compounds. The data demonstrate the serious problem of exact control of stoichiometry and the starting melt composition during the crystal growth from melt. As a result the as-grown crystals are enriched by native point defect contents occasionally exceeding markedly the equilibrium concentration at stoichiometric composition. Therefore, during cooling down, the excess atoms agglomerate in precipitates as second phases due to the retrograde solidus (compare with Fig. 3.13).

Tab. 3.2: Mole fraction position of congruent melting point x_B(cmp) and total pressure minimum $x_B(p_{min})$ relative to the stoichiometric composition $x = 0.5$ in selected semiconductor compounds (*with permission of Elsevier, Springer Nature, Research Signpost, and by open access of IntechOpen*).

Compound	ZnSe	ZnTe	CdS	CdTe	GaAs	InP	PbTe	SnTe
CMP (°C)	1522	1300	1475	1092	1238	1062	925	1063
x_B (CMP)	0.501	0.5001	0.495	0.50001	0.50008	0.49991	0.5002	0.504
p_{min} (MPa)	0.053	0.064	0.220	0.023	~10^{-10}	~10^{-11}	~10^{-11}	0.07
x_B (p_{min})	0.513	0.507		0.507	~0	~0	~0	~0.504

In all listed compounds, the stoichiometric composition is $x = 0.5$.
Data from: H. Wenzl et al.J. Crystal Growth109 (1991) 191; R. Brebrick, Comp. Phase Diagr. Thermochem. 34 (2010) 434; N. Yellin, S. Szapiro, J. Crystal Growth 73 (1985) 77; J. Greeberg, Thermodynamic Basis of Crystal Growth (Springer 2002); V.L. Zlomanov, A.V. Novoselova, p-T-X-diagramy sostojanija sistem metal-chalkogen (Nauka, Moskva 1987); P. Rudolph in M. Isshiki (ed.), Recent Development of Bulk Crystal Growth, Ch.5 (Res. Signpost 1998) p. 127; E. Rogacheva in: A. Innocenti, N. Kamarulzaman (eds.), Stoichiometry and Material Science, Ch. 5 (Intech 2012) p. 105.

The *incongruent evaporation* does shift the melt composition away from the weighed near-stoichiometric feedstock even in evacuated and sealed ampoules, as they are used in Bridgman growth configurations. Usually, the pieces of the solid starting charge cannot completely fill the whole ampoule volume. Consequently, during the melting process, the feedstock slumps down, leaving an empty vacuum space over the molten material (Fig. 3.17). Next, the melt evaporates incongruently fills this space by the most volatile gas component (e.g., the gas phase of CdTe near its melting point is constituted of 99.6% of Cd species). As a result, the number of atoms of the high-volatile component evaporating into the vapor phase leaves an equal number of nonvolatile atoms within the melt in excess. Using the ideal gas equation, the excess concentration N_{ex} per cm^3 can be estimated as

$$N_{ex} = \frac{p_A N_A}{RT} \frac{V_V}{V_L} \tag{3.63}$$

where p_A is the partial pressure of the high-volatile component A, N_A the Avogadro's constant, R the universal gas constant, T the absolute temperature, and V_V/V_L the ratio of the volumes of the melt and free gas inside an ampoule. At constant cross section, the V_V/V_L ratio can be replaced by h_V/h_L with the heights h_V and h_L being of melt and gas sections, respectively. To go back to the example of molten CdTe in an evacuated ampoule with constant diameter at temperature $T = 1100$ °C, i.e., some degree above

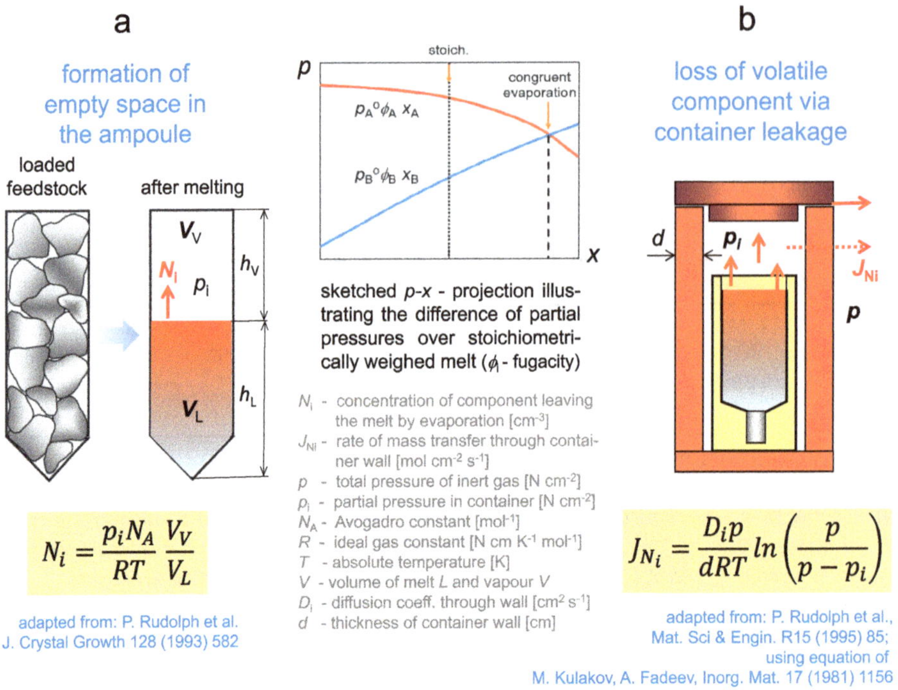

a
formation of empty space in the ampoule
loaded feedstock
after melting
V_V
N_i
p_i
h_V
V_L
h_L
$N_i = \frac{p_i N_A}{RT} \frac{V_V}{V_L}$
adapted from: P. Rudolph et al. J. Crystal Growth 128 (1993) 582
stoich.
p
congruent evaporation
$p_A^o \phi_A x_A$
$p_B^o \phi_B x_B$
x
sketched p-x - projection illustrating the difference of partial pressures over stoichiometrically weighed melt (ϕ_i - fugacity)
N_i - concentration of component leaving the melt by evaporation [cm^{-3}]
J_{Ni} - rate of mass transfer through container wall [mol cm^{-2} s^{-1}]
p - total pressure of inert gas [N cm^{-2}]
p_i - partial pressure in container [N cm^{-2}]
N_A - Avogadro constant [mol^{-1}]
R - ideal gas constant [N cm K^{-1} mol^{-1}]
T - absolute temperature [K]
V - volume of melt L and vapour V
D_i - diffusion coeff. through wall [cm^2 s^{-1}]
d - thickness of container wall [cm]
b
loss of volatile component via container leakage
d
p_i
J_{Ni}
p
$J_{N_i} = \frac{D_i p}{dRT} ln\left(\frac{p}{p - p_i}\right)$
adapted from: P. Rudolph et al., Mat. Sci & Engin. R15 (1995) 85; using equation of M. Kulakov, A. Fadeev, Inorg. Mat. 17 (1981) 1156

Fig. 3.17: Two cases of directional solidification of compounds with incongruent evaporation, leading to the drift away from stoichiometric composition due to empty ampoule space (a) and container leakage (b) (*with permission of Elsevier and Izd. Nauka Russ. Feder*).

the congruent melting point, the Cd partial pressure is roughly 5×10^{-2} MPa around two magnitudes of order more than that of tellurium (see detail graph in Fig. 3.16, right). As a result, the free ampoule volume is enriched by the vaporized Cd atoms. Thus, using eq. (3.63), the resulting residual Te excess in the melt is of the order of $N_{ex} \approx 5 \times 10^{18}\, h_V/h_L$ cm^{-3}. In practice, the values of the height ratios lie between 0.3 and 0.2, which results in Te excess concentrations in the melt volume and, thus, at the beginning in the growing CdTe crystal of about 10^{18} to 10^{19} cm^{-3}. Due to the continuous reduction of h_L during the directional solidification, a further slight increase of the Te excess concentration can be expected as long as the partial Cd pressure does not drop with increasing deviation from stoichiometry. It should also be mentioned that the effect of segregation during directional solidification promotes an additional Te enrichment in the melt, caused by the less tellurium solubility in the solidifying CdTe compound than in its melt (see also Fig. 3.13).

An additional situation of component loss is illustrated in Fig. 3.17 (right). When sealed ampoules are replaced by covered containers, the most volatile component can leave the gas volume over the melt by diffusion through small gaps and the container wall. The rate of mass transfer J_{Ni} through a porous container wall (e.g., made of graphite) can be estimated by the added relation.

Summarizing, the incongruent evaporation proves to be one of the general thermodynamic problems at crystal growth of multicomponent materials. Therefore, the finding out of proper measures for in situ stoichiometry control turns out to be a cardinal action of crystal growth mastering. Roughly approximated, in Bridgman/VGF growth, a simple measure is the ab initio admixture of the excess content of the dissociating component to the feedstock material by using eq. (3.63). However, the maintenance of a quasi-steady near-stoichiometric situation during the whole growth process proves to be a much greater challenge.

3.2.4.2.3 The control of stoichiometry via in situ vapor–liquid–solid-phase equilibration

Compounds of exact stoichiometric composition show excellent, even singular, physical properties due to the minimum intrinsic point defect and dislocation content. Therefore, it is a long-term goal to grow such crystals of extraordinary qualities. However, it proves to be a big technological challenge, above all due to the complexity of exact feedstock weighing (see Fig. 3.17) as well as the effect of segregation outside the congruent melting point, where the phase equilibrium splits into liquidus and solidus (see Section 3.2.4.2.2). Further, as is shown in Fig. 3.14, the CMP does not coincide with stoichiometry. Thus, the growth of compound crystals with exact stoichiometric composition from the melt requires a certain off-stoichiometric melt composition to meet the intersection point of solidus with the stoichiometric composition, as is sketched by the left crystallization path in Fig. 3.13 (Cd-rich for CdTe, Ga-rich for GaAs, and Nb_2O_3-rich for $LiNbO_3$, for example). The difficulty is now to keep constant the related melt composition during the whole crystallization process. As we demonstrated in Section 3.2.4.2.2, this proves to be impossible in growth arrangements with quasi open or leaking containers and, even in closed ampoules without any counteraction.

Actually, Section 3.2.4.2.2 mentioned both variants of (i) liquid encapsulation of the melt surface and (ii) intentionally off-stoichiometric feedstock weighing as applied to the growth of semiconductor compounds by the liquid-encapsulated Czochralski (LEC), vertical Bridgman (VB), and vertical gradient freeze (VGF) techniques. However, the most precise control of stoichiometry proves to be the steady in situ maintenance of the equilibrium vapor phase pressure over the liquid phase, generated by an extra source of the most volatile component, which is located at the appropriate vaporization temperature. For that, the study and linkage of the T–x and p–T projections arises as an essential thermodynamic precondition. Of course, one should have the most detailed possible information about the phase diagram of the given material system. In the following, such a handling will be demonstrated by an example.

Figure 3.18 shows the p_{As}–T and T–x projections of the system Ga–As, whereby p_{As} in Fig. 3.18a presents the total arsenic pressure, being the sum of the partial pressures of the dominating species As_2 and As_4 (V, L and S stand for vapor, liquid, and solid). The homogeneity range of the compound GaAs, according to the newest calculations and experimental experiences, is asymmetric with regard to the stoichiometric composition and completely deviated toward As-rich concentrations (Fig. 3.18b). Also, the CMP at 1238 °C is located on the As-rich side, in equilibrium with a vapor pressure of ~0.2 MPa (2 atm). Thus, in order to ensure the crystal growth near the CMP composition of $x_{As} \approx 0.50008$, the vapor phase pressure above the melt must be kept constant at 0.2 MPa. From the p_{As}–T projection, it follows that such a pressure corresponds to the arsenic temperature $T_{As} \approx 610$ °C on the $S_{As} + V_{As}$ curve (arrow 1). On the other hand, the "left" solidus arm of the existence region of GaAs meets the stoichiometry at about 1150 °C, corresponding with the liquidus mole fraction of $x_L \approx 0.25$. The translation into the p_{As}–T diagram results in a corresponding arsenic equilibrium pressure of $p_{As} \approx 10^{-2}$ MPa at $T_{As} \approx$ 500 °C (arrow 2), suggesting growth near stoichiometric GaAs crystals. All higher arsenic pressures would lead to an arsenic excess δx_{As} within the GaAs compound. At the same time, at all parts inside the growth container, the temperature must be kept higher than T_{As} to avoid condensation on the walls. Thus, by varying the arsenic pressure in the enclosure, the composition of the melt and hence that of the solid can be fixed.

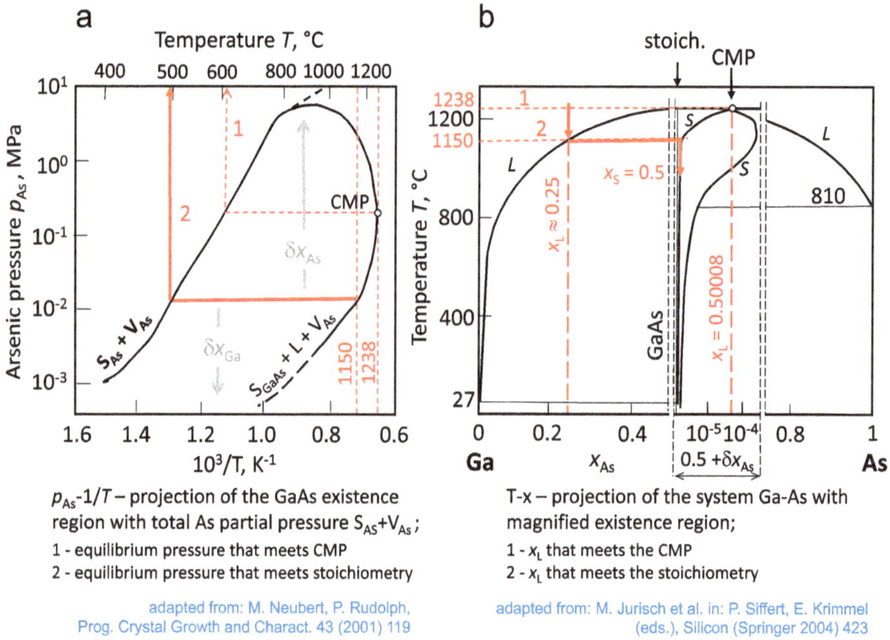

Fig. 3.18: Combination of p–T (a) and T–x (b) projections of the system Ga–As for estimation of the equilibrating pressure and temperature of an As extra source to obtain stoichiometric GaAs (*with permission of Elsevier (a) and of Springer Nature (b)*).

Today, growth from a nonstoichiometric As-rich melt is preferred for a number of reasons, such as generation of the mid-gap donor EL2 (As_{Ga} antisite) as a prerequisite for semi-insulating properties, avoidance of twin formation during crystal growth, and better homogeneity of electrical properties. However, as was analyzed for some other compounds, a near-stoichiometric growth also offers certain advantages, such as minimization of the content of intrinsic point defects, second-phase particles, dislocation density, and dislocation cell patterns. The main methodical problem of stoichiometric GaAs growth proves to be the correspondingly marked excess of gallium in the melt that would require an effective artificial mixing to avoid constitutional supercooling at the growing interface. More details will be given in the lecture parts III and IV.

It is noteworthy that an exact three-phase equilibrium according to the above phase projections can only be obtained when both the solid phase boundary of the growing crystal and the adjacent melt simultaneously and contact the vapor phase. Such melt growth methods are horizontal Bridgman (HB) technique and horizontal zone melting. In fact, the first melt growth experiments with vapor equilibration were carried out with semiconductor compounds (CdTe, GaAs, InAs, GaP, etc.) by HB arrangements, sketched in Fig. 3.19a. In this method, the shoulder-free crystal, melt and, thus, also the growing melt–solid interface are in direct contact with the equilibrating vapor phase generated by an extra source of volatile element (see Fig. 3.19a). Whereas the growth boat is placed in a selected temperature gradient near to the *AB* melting point, named high-temperature heating region, the source of the volatile component *A* or *B* is located within a low-temperature heating section, to deliver the equilibrium pressure at the appropriate temperature. Due to the direct contact with the vapor phase, the required liquid concentration is situated right next to the melt–solid interface. In addition, the direct contact between the saturated vapor and the as-grown free crystal surface seems to be an effective condition for in situ ingot annealing within the atmosphere of excess component.

As compared to this, in the vertical melt-growth arrangements of unidirectional crystallization, like the VB, VGF, and Kyropoulus techniques, the situation is somewhat differing due to the missing direct contact of the melt–solid interface with the gas atmosphere (see Fig. 3.19b). In this case, the crystallization front is totally covered by the melt column. That means, to obtain the required melt composition at the growing interface, the vapor pressure must be adapted to the local *A*:*B* relation, depending on the degree of vapor–liquid separation at the melt surface, the diffusion- and convection-driven concentration level within the melt, and the melt–solid segregation at the propagating interface. As is well known, the segregation effect causes an enrichment of the excess component in the form of a diffusion boundary layer, the height and width of which are dependent on the growth velocity and stirring degree of the melt (see lecture part III). The related concentration profile is sketched in Fig. 3.19b. A quasi-balanced situation is given by the functional dependence between the growth velocity v and the above listed parameters as

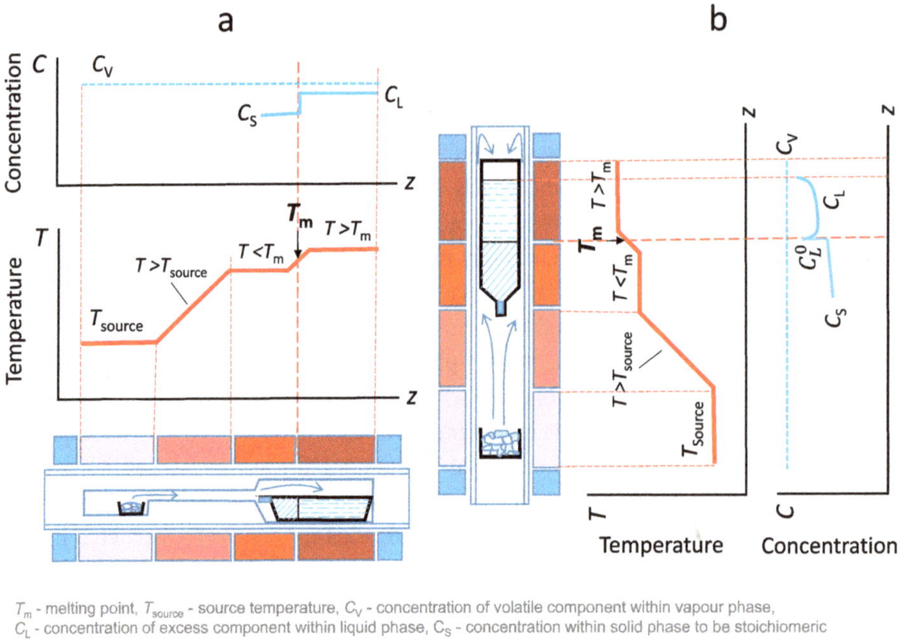

Fig. 3.19: Schemes of horizontal (a) and vertical (b) Bridgman/VGF growth of compounds with equilibrating extra source to control near-stoichiometric crystal composition. The axial temperature and concentration profiles are added.

$$v \cong \frac{D_L}{\delta_D}\left(\frac{c_{\text{ex}}^V k_{\text{ex}}^{VL}}{c_{\text{ex}}^{L0}} - 1\right) \tag{3.64}$$

where D_L is the diffusion coefficient of the excess species within the melt, δ_D the thickness of diffusion boundary layer at the melt–solid interface, c_{ex}^V the concentration of excess component in the vapor phase above the melt surface delivered by the extra source, k_{ex}^{VL} the separation coefficient of the excess component at the melt–vapor interface, and c_{ex}^{L0} the concentration of the excess component directly at the growing melt–solid interface with location coordinate $z = 0$ (derivation of eq. (3.64) is explicated in the *Spec box 3.7*). It must be pointed out that for obtaining a stoichiometric composition in the growing crystal, the value c_{ex}^{L0} is fixed by the desired liquidus-solidus balance according to the given T–x phase projection so that $c_{\text{ex}}^S = k_{\text{ex}}^{Ls} c_{\text{ex}}^{L0}$ with k_{ex}^{Ls} the segregation coefficient of the excess component usually being significantly less than unity due to the large difference between liquidus und solidus along the boundaries of the existence region (see Fig. 3.14).

Spec box 3.7: Growth velocity at stoichiometry-controlled vertical Bridgman technique

In order to achieve the exact stoichiometric composition of a growing crystalline compound *AB*, there should be no any *A* or *B* excess concentration in the solid so that $c_{ex}^{S} \equiv \delta x = 0$. Considering the effect of segregation, the excess concentration *A* or *B* in the melt directly at the interface with current location coordinate $z = 0$ must be c_{ex}^{L0}, in agreement with the balanced liquidus point in the phase diagram (see Figs. 3.13 and 3.18). That means the melt–solid segregation coefficient of the excess component becomes $k_{ex}^{LS} = c_{ex}^{S}/c_{ex}^{L0}$ (in mole fraction $= \delta x_{ex}^{S}/\delta x_{ex}^{L}$, with δ the deviation from stoichiometry $x_{AS} = x_{BS} = 0.5$). For a striven stoichiometry, the value δx_{BL} must agree with the liquidus position, equating the stoichiometric solidus value $\delta x_{BS} = 0$.

As is well known, at a propagating front of crystallization, a *diffusion boundary layer* is formed. That means, the excess component *A* or *B* is enriched in front of melt–solid interface in the form of an exponentially sloped course. Thus, the value of c_{ex}^{L0} denotes the height of the diffusion boundary layer directly at the interface. Its thickness δ_D is determined by the degree of convection in the melt. Its estimation as a function of the mass transport regime within the fluid phase is described in many textbooks. Due to the complexity of the exact definition, it is mostly used as a variable range of functions. In order to estimate the height and, therefore, the stoichiometry in determining value of c_{ex}^{L0}, we use the following relation from textbooks:

$$c_{ex}^{L0} \cong \frac{c_{ex}^{L}}{\left[k_{ex}^{LS} + (1 - k_{ex}^{LS})\exp\left(-\frac{v\delta_D}{D_L}\right)\right]} \tag{B3.7-1}$$

where c_{ex}^{L} is the mean concentration of the excess component within the melt far away from the diffusion boundary layer, v the crystallization velocity, and D_L the diffusion coefficient of the excess component in the melt. If we look at the solidus course of any magnified existence region, such as in Fig. 3.14, it is immediately striking that the slope is significantly deeper than that of the liquidus. As a result, the segregation coefficient k_{ex}^{LS} is much less than unity and can be neglected. Therefore, at such first approximation, eq. (B3.7-1) is simplified as

$$c_{ex}^{L0} \cong c_{ex}^{L} \exp\left(\frac{v\delta_D}{D_L}\right) \tag{B3.7-2}$$

Practically, the amount of c_{ex}^{L} is determined by the weighing of the feedstock and nearly equivalent with mole fraction of $x_L \approx 0.5$ (note, the accurate value is then adjusted by the partial antipressure of the extra source during the growth process). To find out this pressure for a VB arrangement, one must estimate the *V*–*L*–*S* chronology of the concentration as sketched in Fig. 3.19b. First, the composition difference between the melt and vapor is given by the separation ratio $k_{VL} = c_{ex}^{L}/c_{ex}^{V}$ Assuming *B* as excess component becomes $k_B^{VL} = c_B^L/c_B^V \equiv x_B^L/x_B^V$. In the case of an ideal mixed material system with two components *A* and *B*, the concentration in mole fraction of *B* evaporated in the free volume can be expressed from Raoult's relation as

$$x_B^V = \frac{p_B}{p_B + p_A} = \frac{p_B}{p_{tot}} \tag{B3.7-3}$$

with the partial pressures p_B being calculable by using the empirically obtained relation $\ln p_B = \ln C - \Delta H_v/RT$, where *C* is an experimental constant, ΔH_v the heat of vaporization of the liquid, and *R* the universal gas constant. Note that the gas phase often consists of vapor species in molecular form, such as As_2 and As_4, for example. In such a case, the notation of the partial pressure in eq. (B3.7-3) must be replaced by np_{B_n}, with *n* being the number of atoms that form a molecular species.

Assuming that the vapor phase composes almost dominantly of the species of the volatile component B ($x_B^V \approx 1$) and the melt phase of nearly the half ($x_B^L \approx 0.5$), the relation $k_B^{VL} = x_B^L / x_B^V$ becomes ≈ 0.5. The concentration of excess component in the vapor is roughly estimated via ideal gas rule,

$$c_{ex}^V = \frac{p_{ex} N_A}{TR} \tag{B3.7-4}$$

where $p_{ex} = p_B$ is the partial pressure of the volatile excess component that is produced in the extra source, T the total temperature of the gas phase, N_A the Avogadro's constant, and R the universal gas constant.

Next, the mass balanced crystallization velocity of the propagating melt–solid interface v is related but directed in the opposite direction of the mass flow density of the excess component incorporating into the growing crystal j_S, which is equal to the sum of the mass flow density j_L in the melt directly at the interface and the flow $-j_D$ diffusing back from the enriched diffusion boundary layer into the melt volume, given by the first Fick's law $D_L \, \partial c_{ex}^L / \partial z$ (D_L – diffusion coefficient of excess component in the melt, z – coordinate perpendicular to the growing interface)

$$j_S = j_L + (-j_D) \tag{B3.7-5}$$

$$c_{ex}^S(-v) = c_{ex}^{L0}(-v) - D_L \frac{\partial c_{ex}^L}{\partial z} \tag{B3.7-6}$$

$$v\left(c_{ex}^S - c_{ex}^{L0}\right) = D_L \frac{\partial c_{ex}^L}{\partial z} \tag{B3.7-7}$$

where c_{ex}^{L0} is from eq. (B3.7-1). By activating the equation toward v, the mass balanced crystallization rate is given by

$$v = \frac{D_L}{\left(c_{ex}^S - c_{ex}^{L0}\right)} \frac{\partial c_{ex}^L}{\partial z} \tag{B3.7-8}$$

Replacing c_{ex}^S by $k_{ex}^{LS} c_{ex}^{L0}$ and assuming the concentration gradient at the interface to be a near constant, the derivative can be expressed as

$$\frac{\partial c_{ex}^L}{\partial z} \approx \frac{c_{ex}^{L0} - c_{ex}^L}{\delta_D} \tag{B3.7-9}$$

Setting all relations and approximations from above, i.e., $k_{ex}^{LS} - 1 \approx -1$ and $c_{ex}^L = k_{ex}^{VL} c_{ex}^V$, after some arithmetic, it becomes the growth velocity responsible for near-stoichiometric crystallization as

$$v = \frac{D_L}{\delta_D}\left(\frac{k_{ex}^{VL} c_{ex}^V}{c_{ex}^{L0}} - 1\right) \tag{B3.7-10}$$

In order to prevent *thermal and morphological instability* at the growing melt–solid interface in a given temperature gradient, the crystallization velocity must be fixed at a constant undercritical value (see lecture part III) so that eq. (3.64), respectively, (B3.7-10) can be adjusted toward c_{ex}^V as

$$c_{ex}^V = \frac{c_{ex}^{L0}}{k_{ex}^{VL}}\left(\frac{v\delta_D}{D_L} + 1\right) \tag{3.65}$$

Usually, due to a favorable low temperature gradient at VB and VGF growth, the crystallization velocity v has to be relatively low, in the range of 1–5 mm/h. The diffusion coefficient in melt D_L is around 10^{-5} cm^2/s. The thickness of the diffusion boundary layer δ_D depends on the mixing (convection) level in the melt, and is of the order of some μm. Thus, the term within the bracket becomes nearly unity and eq. (3.65) can be approximated as $c_{ex}^V \approx c_{ex}^{L_0}/k_{ex}^{VL}$.

For practical reasons, it is reasonable to know the temperature of the extra source, which can be estimated from the general temperature dependence of the partial pressure around the stoichiometric composition $\ln p_i = C_1 - C_2/T$ (taken from tabular material values) by linking it with the total pressure over the melt p_{tot} as

$$T_{ex} = \frac{C_2}{C_1 - \lg\left[\frac{\alpha x_L}{1 + x_L(\alpha - 1)} p_{tot}\right]} \tag{3.66}$$

where C_1 and C_2 are material constants and α is the relative volatility of the partial pressures of all components presented in the given material system. The detailed derivation is given in *Spec box 3.8*.

The value of p_{tot} depends on the temperature at the melt–vapor surface T_{surf}, which can be estimated for a two-component system A–B from the *Clausius–Clapeyron* equation as

$$p_{AB}^{tot} = p_{AB}^{T_m} \exp\left[-\frac{\Delta H_V}{R}\left(\frac{1}{T_{surf}} - \frac{1}{T_m}\right)\right] \tag{3.67}$$

where $p_{AB}^{T_m}$ is the pressure of the AB compound at its melting point T_m and ΔH_v the enthalpy of vaporization (see *Spec box 3.8*). The value of ΔH_v can be estimated from the slope of the congruent sublimation line ($S = L$) in the given p–T coordinates.

As an example, let us assume the growth of near-stoichiometric CdTe crystals by the VB method with a Cd extra source (in reality, such experiments have been performed by the author, together with his former team, and described in publications). The scheme of the growth arrangement and related axial temperature distribution is sketched in Fig. 3.20a.

It is noticeable that in contrast to Fig. 3.19, no temperature plateau along the melt column but an increasing temperature course with a gradient of 8 K/cm was presented. As a result, the melt surface temperature T_{surf} was continuously decreasing during the crystallization process due to the upward motion of the heater arrangement. In accordance with this, the temperature of the Cd extra source $T_{ex} \equiv T_{Cd}$ was also decreasing with nearly the same cooling rate as the melt surface. At the same time, the starting values of T_{Cd} was determined by the positioning of the source reservoirs (their distance from the bottom of the growth ampoule), sketched in Fig. 3.20a by the two positions 2 and 3. For the calculation of T_{Cd} cooling down from 1200 °C to 1092 °C, eqs. (3.66) and (3.67) and the following quantities are used: $\lg p_{Cd}(\text{atm}) = 5.119 - 5,317/T$, $\lg p_{Te_2}(\text{atm}) = 4.719 - 5,960/T$, $\alpha = p_{Cd}/p_{Te_2} \approx 2.5$ (at ~1100 °C), $\Delta H_v = 198.45\,\text{kJ/mol}$, $p_{AB}^{T_m} = p_{CdTe}^{T_m} = 2.5$

atm. The mole fraction of the Cd-rich liquidus position that meets the stoichiometric composition of the solid CdTe phase is $x_B^L = x_{Cd}^L = 0.49985$, equaling a Cd excess of $\delta x_{Cd} = 1.5 \times 10^{-4}$ (see Fig. 3.14b).

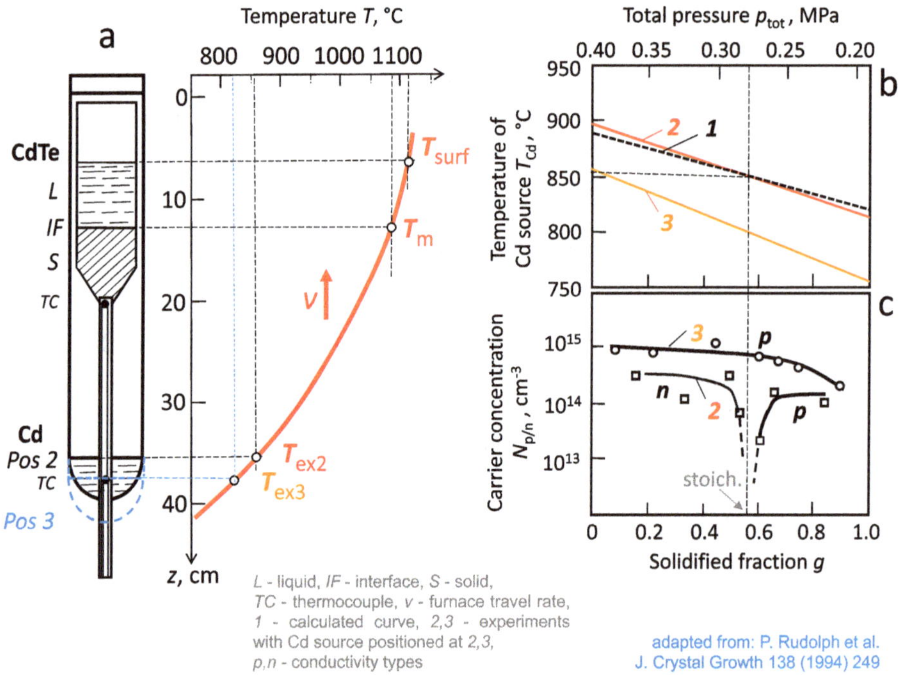

Fig. 3.20: Schematic arrangement of vertical Bridgman growth of CdTe with Cd extra source to study the conditions for stoichiometric crystals (a), the required Cd source temperatures (b) and the related measured free carrier concentration along the solidified fraction (c) (*with permission of Elsevier*).

Spec box 3.8: Extra source parametrization at stoichiometry-controlled melt growth

Assuming a binary system of ideal mixed melt *A*–*B*, the vapor pressure of which follows the Raoult's law. In order to obtain the equilibrium distribution of the most volatile component *B* between the melt and the vapor, we adopt the Rayleigh equation for fractional distillation of ideal melt mixtures

$$\frac{x_B^V}{x_A^V} = \frac{x_B^V}{1 - x_B^V} = \frac{p_B}{p_A} \frac{x_B^L}{1 - x_B^L} \tag{B3.8-1}$$

where x_B^V and x_A^V are the mole fractions and p_A and p_B the partial pressures of the pure constituents *A* and *B* in the melt (*L*) and vapor (*V*), respectively. Rearranging, eq. (B3.8-1) becomes

$$x_B^V = \frac{\alpha\, x_B^L}{1 + (\alpha - 1) x_B^L} \tag{B3.8-2}$$

where α is the relative volatility p_B/p_A. Then, the partial pressure of *B* in an ideal mixture is given by

$$p_{BV} = p_{AB}^{\text{tot}}\, x_B^V \tag{B3.8-3}$$

with p_{AB}^{tot} the total pressure over the *A*–*B* melt. Substituting eq. (B3.8-2) into (B3.8-3) yields

$$p_{BV} = p_{AB}^{\text{tot}} \frac{\alpha\, x_B^L}{1 + (\alpha - 1) x_B^L} \tag{B3.8-4}$$

The expression for the partial vapor pressure of a given component can be obtained from the Clausius–Clapeyron equation in the general form of

$$\lg p_{BV} = C_1 - C_2/T = C_1 - \frac{\Delta H_V}{RT} \tag{B3.8-5}$$

where ΔH_V is the enthalpy of vaporization, R the universal gas constant, and T the total temperature identical with the temperature of the extra source. C_1 and C_2 are material-specific constants listed in tables. Combining eq. (B3.8-4) with eq. (B3.8-5) leads to the temperature of the extra source of the volatile component *B*

$$T_{\text{ex}} = \frac{C_2}{C_1 - \log\left[\frac{\alpha\, x_B^L}{1 + x_B^L(\alpha - 1)} p_{AB}^{\text{tot}}\right]} \tag{B3.8-6}$$

where p_{AB}^{tot} depends on the temperature at the melt–vapor surface T_{surf}. It can be estimated from the Clausius–Clapeyron equation as

$$p_{AB}^{\text{tot}} = p_{AB}^{T_m} \exp\left[-\frac{\Delta H_V}{R}\left(\frac{1}{T_{\text{surf}}} - \frac{1}{T_m}\right)\right] \tag{B3.8-7}$$

Note, if more than two constituents are presented within the melt, as in pseudo-ternary or quaternary mixed systems (see *Spec box 3.2*), one has to consider the relative volatilities of a given (excess) component with respect to all other components in the melt. $\alpha_B(A) = p_B/p_A$, $\alpha_B(B) = p_B/p_C$, …, $\alpha_B(i) = p_B/p_i$ have to be considered by modifying eqs. (B3.8-1)–(B3.8-7).

The resultant theoretical temperature course of the Cd source to maintain constant stoichiometric composition along the whole crystal as function of the solidified fraction g is shown in Fig. 3.20b by curve 1. Then, two experimental curves 2 and 3 are added. They differ in terms of the starting values of the Cd source, $T_{\text{Cd}} = 895$ °C and 855 °C, respectively. Whereas the experimental curve 2 matches the theoretical program quite well and intersects the calculated curve 1 at $g = 0.6$, the curve 3 is below curve 1 during the whole crystallization process, corresponding to an off-stoichiometric growth of a slightly tellurium-rich crystal (see Fig. 3.20b). In fact, the analysis of the free carrier concentrations along the undoped as-grown crystal with such Cd source program 3 yielded a continuous p-type conductivity, characteristic of a Te-rich point defect situation (Fig. 3.20c). In contrast, at the intersection point between curves 1 and 2, a change from slightly Te- to slightly Cd-rich composition takes place due to the transition from lower to higher T_{Cd}, compared with the calculated course. As a result, the free carrier concentration along the related undoped as-grown CdTe crystal shows a transition from p- to n-type conductivity, very near to this point, corresponding with $T_{\text{Cd}} \approx 850$ °C (Fig. 3.20c). That means, the deviation from stoichiometry in the direction of Cd excess produces free electrons by Cd interstitials but in the direction of Te excess free holes by Cd vacancies. It is worth noting that during our research with VB and HB methods, we observed, in agreement with literature, a characteristic lowered temperature T_{Cd} for the *p-n*-transition point in horizontally grown CdTe crystals. In other words, compared to VB, a slightly lower Cd source temperature is required at HB to obtain near-stoichiometric crystals. Obviously, the direct contact of the melt–solid interface with Cd vapor leads to an immediate effect of stoichiometry balance.

Further, the crystallization velocity v as a function on the Cd concentration in the vapor c_{Cd}^V according to eq. (3.64) was combined with the Cd source temperature T_{Cd} (eq. (3.66)). As parameters were applied: $\Delta H_v = 185\,kJ/mol$ K, $c_B^{L0} = c_{Cd}^{L0} = 5 \times 10^{18} cm^{-3}$, $D_L = 10^{-5}\,cm^2/s$, and $k_{ex}^{VL} = k_{Cd}^{VL} = 0.5$. The related graphic is shown in Fig. 3.21. As a further variable, the width of the diffusion boundary layer δ_D as a measure of the degree of melt mixing (convection level) is added. For the typical Bridgman/VGF crystallization velocities in the range of 1–5 mm/h, the related region has been marked in blue. In order to ensure a near-stoichiometric growth, a Cd source temperature of around 860 °C should be held, as marked by the red region. Of course, if the melt surface temperature is changing during the growth process, a related T_{Cd} program must be adapted, as has been mentioned above. As can be seen, the blue-red intersecting area meets the relative thick diffusion boundary layers to be equated with modest convection.

$$v \cong \frac{D_L}{\delta_D}\left(\frac{c_{Cd}^V\,k_{Cd}^{VL}}{c_{Cd}^{L0}} - 1\right)$$

D_L - diffusion coefficient in the melt (10^{-5} cm^2 s^{-1})
δ_D - thickness of diffusion boundary layer (representing quasi the degree of melt mixing)
k_{Cd}^{VL} - separation coefficient between vapor and melt (0.5)
c_{Cd}^{L0} - Cd concentration at the L-S interface to obtain stoichiometric crystal (5 x 10^{18} cm^{-3})

adapted from: P. Rudolph
Progr. Crystal Growth Charct. Mat.
29 (1994) 275

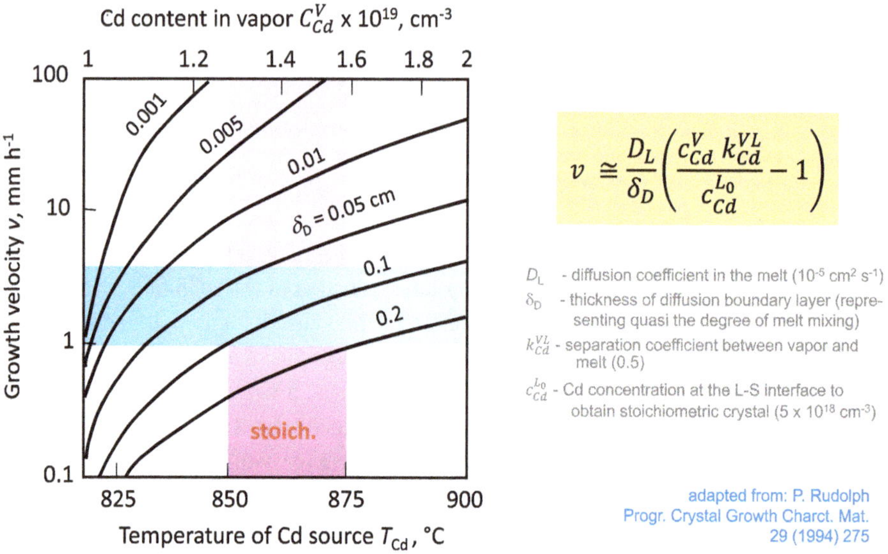

Fig. 3.21: Calculated relation between the growth velocity of CdTe crystals and the temperature of the Cd extra source at different diffusion boundary layer thicknesses (degree of convection) to find out optimum growth conditions for stoichiometric crystal composition, illustrated by the crossed blue (*v*) and red (T_{Cd}) regions (*with permission of Elsevier*).

It is obvious, that these results, demonstrated in Figs. 3.20 and 3.21, underline the importance and capacity of thermodynamic knowledge for well-controlled crystal growth experiments. Of course, identical estimations and experiments can be provided with all other material systems consisting of compounds with existence region too. The obtainment of stoichiometric crystals can improve their electrical, optical, and mechanical parameters. Further, it became clear that the unidirectional crystallization proves to be a dynamic process. Especially during the VGF with nearly constant temperature gradient, the temperature of the melt surface is continuously decreased by computer programming. As a result, the temperature of the extra source T_{ex} has to change synchronously when a stoichiometric crystal composition is aimed. A favored way to master such a challenge should be a *model-based automation*, reviewed by *Winkler* and *Neubert* in the *Handbook of Crystal Growth* Vol. IIB (Elsevier 2015).

3.2.4.2.4 Vapour pressure controlled Czochralski growth without liquid encapsulant

In principle, the Czochralski growth shows a fundamental advantage, namely the growing crystal has no any contact with a solid container wall. This eliminates the risk of foreign nucleation with subsequent polycrystallinity and prevents also contamination by undesired impurities. Without question, therefore, this method is unchallenged for elementary materials, such as silicon and germanium, as well as for hardly dissociating compounds, such as a lot of dielectrics. For years, this method was primarily used to compound semiconductors too, especially the III–Vs, such as GaAs, InP, GaP. However, in order to prevent their strongly pronounced dissociation in the molten state (see e.g. Fig. 3.16) the melt surface had to be covered with the liquid encapsulant B_2O_3. Additionally, the free space above was filled with an inert gas under high pressure. This technique is well-known as liquid encapsulated Czochralski (LEC). Unfortunately, at such a growing arrangement no low-temperature gradients with homogeneous axial temperature distribution and, therefore, no thermomechanical stress- and dislocation-reduced crystal growth is possible. The reason for that is the lacking heat protection of the growing crystal if it emerges from the encapsulant into the furnace atmosphere. In this case a selective evaporation of the volatile component from the too hot crystal surface takes place leading to various defect formations. As a consequence, the LEC is largely replaced by the low-temperature-gradient Bridgman or VGF technique, which is today used for the growth of numerous dissociating compounds and mixed systems, such as GaAs, InP, CaF_2, $Cd_{1-x}Zn_xTe$, $(Ga_{1-x}Sc_x)_2O_3$ with crystal diameters even to 8- and 10-inch like in the case of GaAs and CaF_2, respectively. To minimize the above-mentioned contact between the growing crystal and the container wall, at III–V growth a thin boric oxide skin is applied in between. However, this preserves an old problem, namely the contamination with boron and oxygen, quite comparable to LEC. Furthermore, the widely applied non-stoichiometric growth condition from As-rich melt generates As interstitials precipitating in the as-grown GaAs crystal during the cooling process which are harmful when their size exceeds a critical value (discussed in chs. 3.2.4.2.1 and 5.4; explained by Fig. 3.13). Likewise at the Bridgman or VGF growth of InP from P-rich side phosphor gas bubbles are incorporated forming deleterious microvoids during cooling down.

Thus, this gives rise to the question of whether contactless Czochralski growth without a liquid encapsulant is possible. The author's former team developed this idea further by applying the VCz without B_2O_3 encapsulant summarized by *Rudolph and Kiessling* (2006). The experiments showed some characteristic advantages, such as contactless crystal pulling, direct control of the melt composition (and, thus, crystal stoichiometry) by an extra source of the volatile component, removal of the stress maximum at the interface with the encapsulant, and adjustment of a small axial temperature gradient. Although the following comments may be of interest for quite a lot of compounds, here only exemplary the growth of stoichiometric GaAs is reported. The relationship to thermodynamics is obvious.

Fig. 3.22a shows the VCz arrangement without boric oxide encapsulant schematically. According the *p-T-x* phase relations of GaAs, shown in Fig. 3.18, the composition control was possible due to the direct contact between the As-rich vapor phase and melt surface. The mole fraction of the melt was controlled in situ by the partial arsenic pressure adjusted via the temperature of the As source. The axial temperature gradient at the growing solid–liquid interface was 20 K cm^{-1}. To prevent possible Ga inclusions due to the Ga enrichment in the diffusion boundary layer, all crystals were grown by uncritical pulling rates between 3 and 5 mm h^{-1} depending on the melt composition. Seed and crucible were opposing rotated.

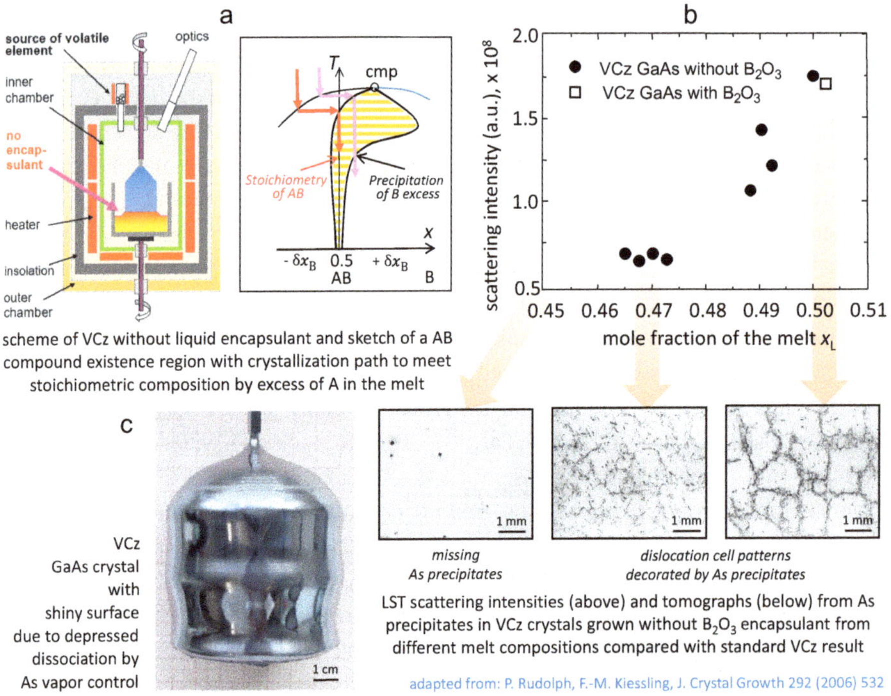

Fig. 3.22: Vapor pressure controlled Czochralski growth without B_2O_3 encapsulant demonstrated by the example of GaAs (*author's sketches (a) and image (c); with permission of Elsevier (b)*).

In 3-inch GaAs crystals (Fig. 3.22c) the following qualities have been obtained: - no twinning, - near-stoichiometric solid composition with markedly reduced As precipitates and without Ga inclusions, - low and radial homogeneously distributed dislocation density ≤ 10^4 cm^{-2}, - reduced boron as low as 10^{15} cm^{-3}, - no oxygen defects and related complexes, - controlled carbon concentration down to 1 x 10^{15} cm^{-3}, - semi-insulating behavior of n-type conductivity with electrical resistivity of > 10^8 Ωcm. One of the targets was the prevention of arsenic precipitates and inclusions which are observed in all standard LEC and VGF grown crystals from As-rich melt. Fig. 3.22b demonstrates the results of the second-phase particle analyses by IR laser scattering tomography (LST) versus the mole fraction of the melt x_L indicating the degree of As precipitate decoration on presented dislocations. As can be seen, the precipitated arsenic excess decreases with enrichment of Ga proportion in the melt. At x_L = 0.45 almost no As excess was detected anymore. In addition, the formation of dislocation cell patterns seems to have depressed, since As interstitials contributing to their formation by necessary climb processes are missing.

Comparing this experimental fact with the *p-T-x* projections in Fig. 3.18, it is apparent that a stoichiometric composition was achieved with a lower excess of Ga in the melt than sketched in the *T-x* diagram. This is probably due to the not yet precisely knowledge of the course of the solidus curve along the existential region. It must therefore be assumed that its coincidence with the stoichiometry line already takes place above 1150 °C. Thus we see how important technological investigations are also for the exact determination of thermodynamic material parameters.

3.2.4.3 Systems with incongruent melting compounds

A number of important material systems, especially multicomponent ones (mostly oxides), contain compounds with *incongruent melting* behavior (compared with compounds of congruent melting discussed in Section 3.2.4.2.1). In most cases, such compounds are of particular significance for lasers, energy converters, high-T_c-superconductivity, fitting substrates in optoelectronic, and high-power devices, evoking great interest to grow them as single crystals from the melt. However, the realization proves to be rather difficult. Compounds of incongruent melting are formed when the interaction energy between the constituents is reduced, compared to those of congruent melting. As a result, their melting points and stabilities are decreased, which leads to a premature decomposition before melting. From the thermodynamic point of view, the Gibbs free energy is lower for one of the compound component and, thus, heating causes it to transform to liquid phase and the more stable another solid phase named *peritectic reaction*. Fig. 3.23a shows the scheme of a *T–x* phase projection demonstrating such a situation. The intermediate compound of incongruent melting, denoted by the stoichiometric composition AB_2, is located between the solid component *B* and the eutectic composition $A + AB_2$. It melts at the *peritectic line*, with the melting point T_m noting the phase intermediate between the solid *B* and liquid *L*. At the peritectic temperature, three phases are in equilibrium, being the maximum for a two-component system. Therefore, according to the Gibbs' phase rule of eq. (3.6), there is no more degree of freedom. To crystallize pure AB_2, its perfect equilibrium with the liquid in the range of $x_{\mathrm{eut}} < x_B < x_{\mathrm{per}}$ must be maintained.

Two possible intermediate compound types are implied: (i) the line type with exact stoichiometric composition (or a very small existence region) along the whole temperature, referred to as *daltonide*, e.g., like AB_2 in Fig. 3.23a, and (ii) the composition with quite pronounced existence region, e.g., $A_{\nu_A \pm \delta x} B_{2\nu_B \mp \delta x}$ (dashed area in Fig. 3.23a) named *berthollide*. Figure 3.23b and c presents two realistic examples. A system with daltonide shown in Fig. 3.23b is the intermediate compound $PrSc_3(BO_3)_4$ with incongruent melting point at 1480 °C on the peritectic line formed between the two quasi-binary compounds $PrBO_3$ and $ScBO_3$. It belongs to the binary borates, the monocrystals of which are known as laser media, allowing doping with lanthanide ions in high concentrations without considerable luminescence during self-quenching. There are many other examples of identical incongruent line-like existence forms having important applicability, such as $SrPrGaO_4$ as substrates for epitaxial growth of high-T_c superconductor thin films, or $Y_3Fe_5O_{12}$ as microwave filters and acoustic transmitters. On the other hand, Fig. 3.23c shows the system Al–Ni, which contains, besides daltonides, the berthollide Al_3Ni_2

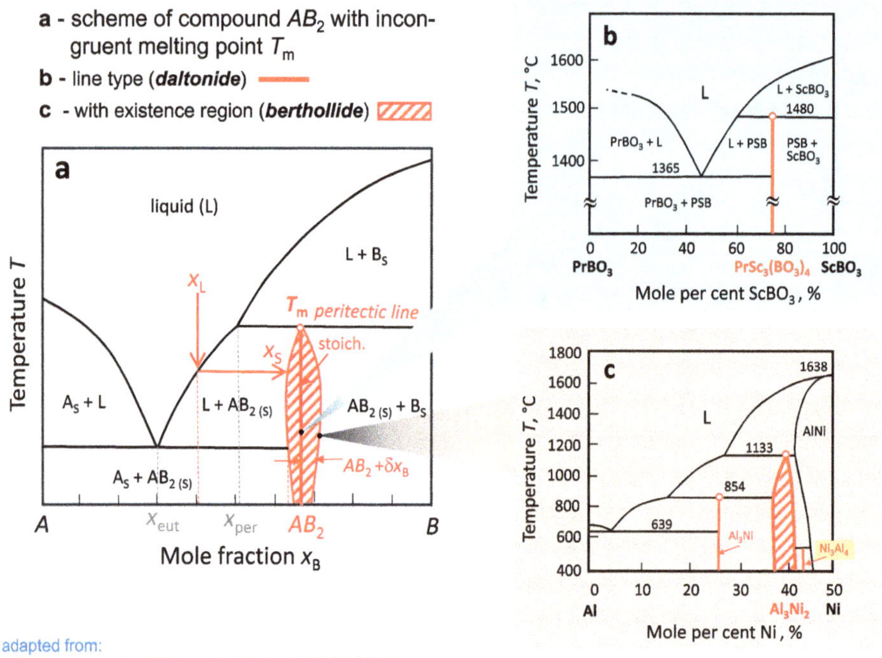

adapted from:
S.T. Durmanov et al.,Optical Materials 18 (2001) 243;
M. Nazarian-Samani and A.R. Kamali, J. Alloys and Comp. 486 (2009) 315

Fig. 3.23: *T*–*x* projections of assumed (a) and real systems (b, c) with intermediate compounds of peritectic decomposition and incongruent melting point (*with permission of Elsevier (b, c)*).

with extremely wide existence region (note, the mole per cents axis is truncated at 50%). Al–Ni composites are used as brazing solder for ultrasonic welding of lithium-ion battery packaging, whereby the intermediate phases can promote or affect the weld quality.

Generally, all compounds with incongruent melting are characterized by the impossibility of their growth directly from a melt of the same composition. Their successful growth is only possible from starting melts with concentrations between the peritectic and eutectic points of the liquidus curve (region $x_{eut} < x_B < x_{per}$ in Fig. 3.23a). Such a melt is quite comparable with a melt–solution possessing a native system component as solvent. Due to the usual enormous composition difference between the solidus and liquidus of peritectic compounds x_S/x_L (in Fig. 3.23 this ratio is ≫1) at unidirectional crystallization, a marked part of the excess component is rejected at the propagating melt–solid interface, causing a high danger of *constitutional supercooling* (see lecture part III). Therefore, often relative low growth rates must be applied. High-quality crystals grown by the top-seeded solution growth (TSSG) have been reported in the related literature. By the way, quite a similar situation occurs at the growth of congruent melting compounds at reduced temperatures, far less than their melting point from a melt with high excess of one of the components acting as solvent. An additional effect of improved axial homogeneity is achieved when a liquid zone of the excess component is used like in the traveling heater method (THM).

Another difficulty concerns the berthollides with extended existence regions. Due to the impossibility of growth from near-stoichiometric melt, their as-grown solid composition is always deviated from stoichiometry toward the liquidus composition (see Fig. 3.23a and c). Furthermore, due to the change of the melt composition from the starting mole fraction $x_L \leq x_{per}$ down to $x_L \geq x_{eut}$, the degree of deviation from stoichiometry follows the given solidus course of the existence region. As a result, such melt-grown crystals are enriched by a marked intrinsic point defect content affecting the electrical, optical, and magnetic parameters. Additionally, the composition of single crystals varies strongly along the growth direction. Of course, for special applications, off-stoichiometric composition may prove favorable. However, when strong stoichiometric crystals are required, post-annealing (of cut wafers) of the vapor of undersaturated component is recommendable.

3.2.4.4 Systems with solid–solid transition

An uncomfortable situation arises when an as-grown crystal, during its cooling down from the crystallization temperature to room temperature, undergoes a solid–solid phase transition, as sketched in Fig. 3.14f. Such transformation from one to another crystalline structure is driven by the principle of Gibbs free energy minimization (see Section 2.1), whereby with reducing temperature, the potential of the low-temperature phase falls below those of the high-temperature one. Hence, a first-order phase transition takes place as was introduced in Section 2.2. and characterized by Tab. 2.1. Due to the reduced kinetic ability, the solid–solid phase transitions proceed often incompletely and are mostly accompanied by generation of structural defects like twins, dislocations, and grain boundaries.

In elementary crystals, the presence of different crystalline structures is known as *allotropism*. In multicomponent materials, it is named *polymorphism*. For instance, diamond and graphite are well-known allotropes of carbon, belonging to the cubic and hexagonal crystal system, respectively (see Fig. 3.4, left). Another example is sulfur, which shows, above 95.6 °C, a monoclinic structure (β-sulfur), but below it, an orthorhombic one (α-sulfur). Further, silicon carbide (SiC) is a well-known semiconductor compound, being unique in this regard due to more than 250 polymorphs. The different polytypes have wide ranging physical properties. Among the SiC polytypes, 6H (hexagonal symmetry) is most easily prepared and best studied, while the 3C (cubic) and 4H polytypes are attracting more attention for their superior electronic properties. Due to its unique conduction band structure, the rhombohedral 9R-SiC may also exhibit improved electron transport properties and could be suitable for high-frequency and high-voltage applications. The polytypism of SiC makes it nontrivial to grow in single-phase state.

There exists an enormous number of publications on allotropism and polymorphism in the fields of material science, and crystal growth being available for more explicit studies. Here, the methodical difficulty will be demonstrated by one example only – the melt growth of ZnSe crystal.

Equal to numerous other II–VI compounds, ZnSe shows dimorphism. At room temperature, the zinc blende structure (ZB) is the stable form showing the stacking sequence aα–bβ–cγ–aα (3C) (Fig. 3.24a). However, at high temperatures, near the congruent melting point $T_{cmp} = 1425$ °C, the wurtzite phase (W) with stacking sequence aα–bβ–aα (2H) does exist (Fig. 3.24b). Therefore, melt-grown crystals pass through a solid–solid phase transition from W to ZB (2H → 3C) during the cooling process. The transition point has been found at about 100 °C below T_{cmp} at 1425 °C. Conversely, a ZB seed crystal would undergo a 3C → 2H transition during heating (Fig. 3.23c). Figure 3.23d sketches a magnified cutout of the T–x phase projection of the system Zn–Se, showing the existence region of ZnSe compound with the W–ZB phase translation. A possible crystallization path from Se-rich melt is added by the red arrows. As can be seen, if the melt of a given mole fraction x_L meet the solidus S_W first, the W phase is crystallized. Then, at $T < 1425$ °C, the solidus S_{ZB} of the ZB phase with composition x_S is reached. Thus, a disassociation into two phases of different composition within a certain temperature interval takes place. Usually, for this process, relatively slow kinetics over a longer period of time is necessary, owing to atomic interdiffusion. Thus, under practical crystal growth conditions, a partial freezing-in of any intermediate state of atomic disorder (e.g., interstitials) and also W rudiments can be expected. However, especially in ZnSe, above 1200 °C, the transformation kinetics proves to be so rapid that remnants of metastable W phase are not observed at room temperature. Considering the high migration mobility, especially of Se atoms, their shift from W-positions to ZB-positions within every second (0001) double net plane is com-

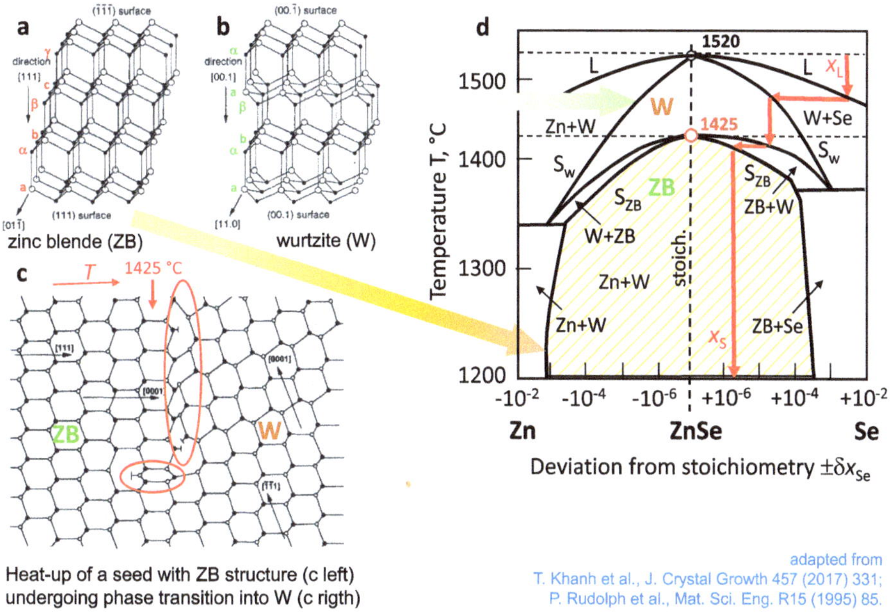

Fig. 3.24: Wurtzite (W)–zinc blende (ZB) solid–solid-phase transition in ZnSe compound (d) leading to crystal structure defects like twinning during cooling of the as-grown crystals from melt and during heat up of a ZB seed (c) (*with permission of Elsevier*).

pleted relatively fast. Owing to the low energy required for this atomic movement, a preferential *layer-by-layer nucleation mechanism* for the ZB phase, parallel to the (0001) plane of the W structures, is most probable. Therefore, a monocrystal with W structure may always translate, on cooling into a single crystalline ZB ingot having one of the {111} planes parallel to the former {0001} plane. In an uniaxial temperature gradient, as is applied at VB or VGF growth during cooling down, the W–ZB transition starts at the coldest region, i.e., at the bottom tip of the as-grown wurtzite crystal. However, due to the very low energetic differences between the 3C and 2H stacking sequences, when more of a certain density of dislocations are already present in the W phase, the generation of *stacking faults* are facilitated. As a result *microtwinned* ZB crystals will be formed, whereby the twinning sequence is dependent on the ZB nucleation mode, which may, in turn, be a function of the degree of supersaturation of the 2H-3C phase translation.

Conversely, by heating up of ZB crystals the situation is far more complicated. Then, at the phase transition point the ZB structure can be transformed into four differently oriented W individuals with only one (0001) plane each. As a result, a W polycrystal containing large angle grain boundaries will be formed (Fig. 3.24c). This fact has to be considered in the growth from melt if a ZB seed is provided. Its transition into the W phase is accompanied by generation of polycrystallinity, making its application quasi impossible. In its place, a wurtzite seed crystal could be used. In principle, the low-temperature growth from melt–solution below the solid–solid transition temperature, e.g., by THM, is favored all the more because the use of seeds of low-temperature ZB phase is allowed.

3.2.5 Ternary systems

With the increasing search for new materials showing specific or improved parameters, suitable for the widening of several application fields, systems with combinations of more than two (or quasi-two) components become more and more important. Apart from the well-known ternary semiconductor mixing systems on the basis of III–Vs or II–VIs, like $In_{1-x}Ga_xAs$, $Ga_{1-x}In_xN$, and $Cd_{1-x}Zn_xTe$, or the long-term experiences on the quasi-ternary system $PbO–Fe_2O_3–Y_2O_3$ for YIG crystallization from PbO solution, currently, the study and epitaxy of diverse newer metallic, semiconductor, oxide, and organic *ternary systems* are under investigation or already applied in industry. For instance, Ni–Al–Ti superalloys are unidirectionally solidified to form turbine blades of high creep and oxidation resistance, $Mg_xTM_{1-x}N$ (TM = Ti, Zr, Hf, Nb) mixed crystals show remarkable optoelectronic properties, $Nd_2O_3–Lu_2O_3–Sc_2O_3$ serves for preparation of perovskite-type mixed crystals, $NdLu_{1-x}Sc_xO_3$, which are used as substrate material for strain engineering of epitaxial perovskite layers, $Li_2O–Na_2O–MoO_3$ component relations are of increasing interest for $LiNa_5Mo_9O_{30}$ crystal production with outstanding luminescent properties, and the ternary organic system DC–NPG–SCN (D-camphor–neopentylglycol–succinonitrile) are used for obtainment of transparent plastic crystals.

The simplest ternary-phase diagram *A–B–C* consists of three binary systems *A–B*, *A–C*, and *B–C* with complete mixing of the liquid and solid phases (*ternary isomorphous system*). But mostly, there are ternary systems of differing characteristics, for instance, when all three binary subsystems show eutectic transitions. Besides this, there are also combined systems composed of one or two binaries with eutectics and one or two binaries of complete mixing, respectively. Further, combinations between systems with peritectics and compound formations are also possible. With the result, the three-dimensional representations of such phase diagrams often become extremely complex. Even today, when the finding of new material qualities on the basis of assembling various proper substance combinations is of increasing importance, an increasing demand for the better geometrical imagination and mastery of ternary-phase diagrams is required. Fortunately, there is a large number of related textbooks, publications, and many excellent internet lectures as well as presentations that are highly recommended. Here, we will give some introductory information only.

Mostly the complex three-dimensional illustrations, shown in Fig. 3.25a, are replaced by horizontal cuts transverse to the vertical temperature *T*-axis, which represent more clearly isothermal compositional planes, like in Fig. 3.25b. The most convenient form is an equilateral triangle. It is clear that three mole fractions of a ternary system *A–B–C* must result in a sum of unity, as $x_A + x_B + x_C = 1$. Thus, there are two independent concentrations. After the Gibbs phase rule [eq. (3.6)] in a system with three components, a maximum of five coexisting phases is possible. However, at fixed low pressure (~0.1 MPa), the variance of the system is decreased and becomes four phases, maximum, in equilibrium. If the temperature T (along the z-axis) is kept free, a three-dimensional T-(x_A, x_B, x_C) – phase projection with concentration axes in form of a triangle base plane (*Gibbs triangle*) can be constructed. Figure 3.25a shows the liquidus planes L of a sketched ternary eutectic system in three-dimensional presentation. Let us start the crystallization path from a given liquidus point Z projected into the two-dimensional plane in Fig. 3.25b. First, only one solid phase (A) is in equilibrium with L, which is obtained by the so-called *primary crystallization*. In the same way as the residual melt is enriched by B and C, the path reaches point Y with two coexisting solid phases (A and C). Finally, the ternary eutectic E (with three coexisting solid phases) of the lowest melting temperature is reached. The dotted lines in the 3D contour represent some isotherms. They also appearing as such in the 2D projection, marked by the given z-level of temperature.

Some of the ternary- and quaternary-phase diagrams show a *miscibility* gap. In these systems, the interaction parameter between two (or more) participating components (or compounds) Ω has a large value > 0 (we introduced Ω in Section 3.2.3; see also Figs. 3.9 and 3.10). Miscibilities are possible in the liquid state and in solid-solutions. An example of insoluble fluids is alcohol (as component A) in water (component B). But, by adding salt (as ternary component C), their solubility increases by separation into an alcohol-rich and water-rich solutions. Miscibility gaps in solid-solutions have been found in quaternary semiconductor III–V systems, for example. Such regions within the phase diagrams exist in $Ga_xIn_{1-x}As_ySb_{1-y}$ or $Ga_xIn_{1-x}P_yAs_{1-y}$, which do not make it easy to produce homogeneous thin films even by liquid-phase epitaxy (LPE). But it has been overcome by tin film epitaxy using very efficient substrate strain-stabilizing effect which reduces the miscibility gap significantly.

Figure 3.25c shows a Gibbs triangle with composition lines of a three-component system A–B–C, in which the pair C–B is partly miscible. First, let us assume an alloy at the black point P within the isothermal section of the completely mixed phase. Then, the three perpendicular distances of P from the sides of the triangle (black dotted lines) correspond to the fractional composition of each alloy partner. This is because, in an equilateral triangle, the sum of the distances of each point add up to the height of the triangle, which is equal to 100% ($x_i = 1$). For an easier reading, one uses the coordination grid added in Fig. 3.25c as thin network. As was mentioned above, two independent variables are needed to characterize a ternary system. Knowing two variables, the third follows, as the remaining from 100% (or the remaining fraction from unity). Turning the focus now to the miscibility gap (red dashed area in Fig. 3.25c). There are two phases in equilibrium. The compositions of the phase relation coexisting at the red point P can be deduced from the *lever rule* along the *tie lines* (red lines) connecting the two phases. Knowing the relative composition A:B:C of the respective line ends L_1 and L_2, each quasi new alloy composition P on that line to be crystallized can be deduced as

$$\%L_1 = \frac{P - L_1}{L_2 - L_1}100\% \text{ and } \%L_2 = \frac{P - L_2}{L_2 - L_1}100\% \qquad (3.68)$$

with $P–L_1$, $P–L_2$, and $L_1–L_2$, the distances between composition P and L_1, P and L_2, and the total tie line lengths $L_1–L_2$, respectively.

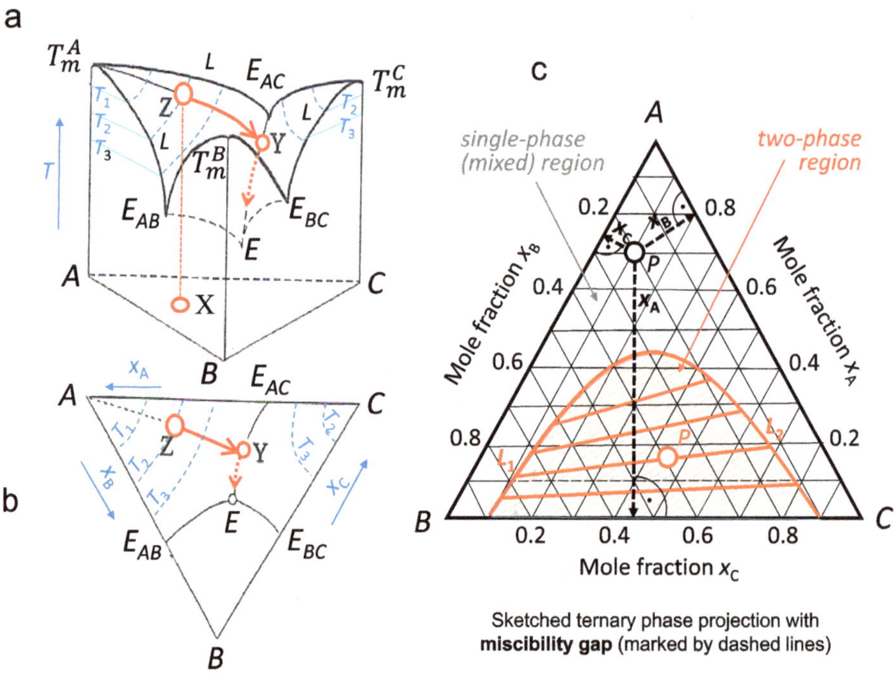

Fig. 3.25: 3D and 2D sketches of ternary-phase diagrams without (a, b) and with two-phase region (c).

Let us demonstrate an example of how to feasibly work with a ternary-phase projection at crystal growth. Fig. 3.26 shows the triangle projection of the system Pb–Sn–Te, important for the growth by THM and LPE of $Pb_{1-x}Sn_xTe$ mixed crystals and epitaxial layers from Te- or Sn-rich melt–solutions. Such crystals and thin films are used in lasers and diodes of IR-wavelengths and recently, also for thermoelectric devices. The system combines the two added binaries with compounds PbTe and SnTe as well as the eutectic system Pb-Sn. Between them, the mixed solid solution $Pb_{1-x}Sn_xTe$ does exist. Some liquidus isotherms are inserted in the Pb–Te–Sn projection as black lines. In order to visualize their origin, the exemplary connections with the related temperature points of 750 °C in the binaries liquidus curves are demonstrated. Also the iso-concentration lines in the mole fraction x_{Sn}^S of the solid solution $Pb_{1-x}Sn_xTe$ are inserted as blue dashed lines. For instance, an aimed solid composition $x_{Sn}^S = 0.2$ can be obtained at the given growth temperature of 600 °C from both Te- or Sn-rich side if the melt–solution compositions are $x^L_{Pb}:x^L_{Sn}:x^L_{Te} = 0.18: 0.07: 0.75$ (point P_1) or 0.40: 0.55: 0.05 (point P_2), respectively. However, contrary to the Sn-rich side (P_2), the Te-rich region (P_1) shows a more ideal behavior. Here, the iso-concentration lines are nearly linear (see Fig. 3.26). That means, at all liquidus temperatures of related Te fraction as solvent, the solid Pb/Sn ratio, determined by the equilibrium distribution coefficient $k_0 = x^S_{Sn}/x^L_{Sn}$, is quasi constant, being of certain methodical advantage.

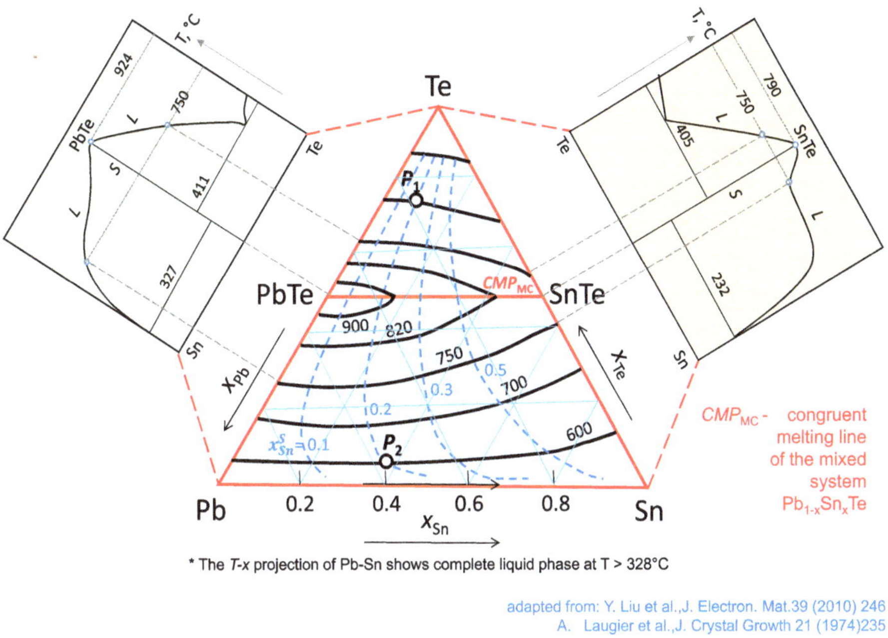

Fig. 3.26: Projection of the liquidus surface by isothermal cuts of the system Pb–Sn–Te (*with permission of Springer Nature (binary-phase diagrams) and Elsevier (ternary-phase projection)*).

3.2.6 The thermodynamic equilibrium segregation coefficient

3.2.6.1 Segregation at melt–solid phase transitions

As has been noted in the previous chapters, the segregation coefficient was already needed to apply occasionally. Let us now turn to the exact thermodynamic determination and application of this very important coefficient affecting markedly crystal growth processes.

The *effect of segregation (or distribution)* is subjected to the concentration difference between solid and fluid phases. Only single component and congruently melting multicomponent systems, treated in Sections 3.1 and 3.2.4.2, respectively, remain almost unchanged during crystallization. In all other systems, the composition of the solid differs from that of the fluid phases (here, the melt). Figuratively, this follows from the difference between the solidus and liquidus, as has been shown above in the various T–x projections. That means, at the nucleation of the solid phase within the molten phase or at a propagating melt–solid interface, a *component redistribution* always takes place. From the beginning, this fact proves to be one of the main challenges in crystal growth. Strictly speaking, because of the impossibility of the existence of absolutely pure elements, segregation does not only occur at crystallization of multicomponent alloys and mixed systems. Even the residual impurities and intentionally doped additives are subjected to this effect. Besides the obtainment of highest purity, the homogeneous distribution of dopants within the crystal belongs to the ab initio mastering of high-quality single crystals.

The diversity of occurrence of segregation is demonstrated by three selected T–x projections in Fig. 3.27a–c. Sketch (a) shows the mixed system K_2NO_3–Na_2CO_3. The figure below shows the magnified K_2NO_3-rich side with red tangents approximating the courses of liquidus and solidus. As is well known, such first approximation is feasible at low concentrations of the related second admixture B (here Na_2CO_3) in order to gauge the equilibrium segregation coefficient $k_0 = x_{BS}/x_{LS}$, being constant at two straight ledges. Of course, in such systems of total mixing, k_0 is continuously changing over the whole range, $0 < x_B < 1$, and the approximation by two tangents must be replaced by a regression (see below). In b) a typical case of very high dilution is shown as in the case of the residual impurity carbon in silicon. From the magnified figure below, it follows that up to the limit of solubility at $x_C \approx 5 \times 10^{-4}$ ($\approx 3.5 \times 10^{18}$ C atoms per cm^3), both liquidus and solidus are well fitted by two tangents so that the constancy of $k_0 = 0.07$ in this region is quite reasonable. Such pronounced effect of segregation makes it relatively easy to purify silicon from carbon but on the other hand, difficult to distribute it uniformly within the as-grown crystal. Finally, (c) shows the quasi-binary system, Nb_2O_5–Li_2O_3, with the stoichiometric compound $LiNbO_3$, having, however, a very wide existence region. As demonstrated in the below sketch, both paths of crystallization, either from Nb_2O_5 - or from Li_2O_3-rich side, are possible to meet the compound composition. However, whereas on the left side liquidus and solidus can be very well approximated by two tangents, on the right side this measure proves to be imprecise, especially concerning solidus. Thus, k_0 is constant on the left (1.04) up to the deviation of the mole fraction from stoichiometry $\delta x \approx -0.1$, but on the right (~0.9), only up to a very small excess of $\delta x = +0.01$, respectively. Also for such a case, especially with a steeper retrograde solidus, we will show a mathematically better approximation below.

Now, let us develop some thermodynamic relations. When a melt of a given composition $[C_{BL}]$, in mole fraction $x_{BL} = [C_{BL}]/([C_{BL}] + [C_{AL}])$, is cooling down to the equilibrium with the solid phase at a given equilibrium temperature T_{eq}, crystallization starts with differing mole fraction $x_{BS} = [C_{BS}]\,/([C_{BS}] + [C_{AS}])$. The correlating *thermodynamic equilibrium (or distribution) coefficient* k_0 results from the equalization of the Gibbs free potentials of solid (S) and liquid (L) phases, as was already performed by eqs. (3.51) and (3.52)

$$\mu_{BS}^{0} + RT\ln x_{BS} + RT\ln \gamma_{BS} = \mu_{BL}^{0} + RT\ln x_{BL} + RT\ln \gamma_{BL} \tag{3.69}$$

and after transforming, it becomes

$$\ln\left(\frac{x_{BS}}{x_{BL}}\right) = \ln k_0 = \frac{\mu_{BL}^{0} - \mu_{BS}^{0} + RT\ln \gamma_{BL} - RT\ln \gamma_{BS}}{RT} \tag{3.70}$$

with the terms of ideal mixing $\mu_{BL}^{0} - \mu_{BS}^{0} = \Delta h_{B}^{0} - T\Delta s_{B}^{0}$ and according eq. (3.38) the excess contributions $RT\ln \gamma_{BL,S} = \Delta h_{BmL,S}^{\mathrm{ex}}$ in the liquid and solid, respectively. Finally, it is

$$lnk_0 = \frac{\Delta h_B^0}{R}\left(\frac{1}{T} - \frac{1}{T_{Bm}}\right) + \frac{\Delta h_{BmL}^{ex} - \Delta h_{BmS}^{ex}}{RT} \tag{3.71}$$

After setting $\Delta s_B{}^0 = \Delta h_B{}^0/T_{Bm}$, with T_{Bm} the melt temperature of the pure added element B, and assuming the excess terms $\Delta h_{BmL,S}^{ex} = 0$ [compare with eq. (3.52)], the thermodynamic equilibrium segregation (or distribution) coefficient for a *system of ideal mixed solutions in melt and solid* is

$$\ln k_{0_{\mathrm{id}}} = \frac{\Delta h_B^0}{R}\left(\frac{1}{T} - \frac{1}{T_{Bm}}\right) \tag{3.72}$$

where Δh_B^0 is the intensive standard enthalpy of the added pure component B, and R the universal gas constant.

As we discussed in Section 3.2.3, in practice, most binary systems show a real behavior of their components. Then, eq. (3.72) for ideal mixing in both phases does not prove to be more satisfactory and the excess terms must be considered. In an usual case of ideal mixing in the melt ($\Delta h_{BmL}^{\mathrm{ex}} = 0$) but *real mixing in the crystalline phase* ($\Delta h_{BmS}^{\mathrm{ex}} \neq 0$), the segregation coefficient is

$$\ln k_0 = \frac{\Delta h_B^0}{R}\left(\frac{1}{T} - \frac{1}{T_{Bm}}\right) - \frac{\Delta h_{BmS}^{\mathrm{ex}}}{RT} \tag{3.73}$$

As mentioned in Section 3.2.3, the determination of the excess enthalpy via the activity coefficient γ proves to be often difficult due to the lack of exact interchange energies between the constituents within the phases. Using the activity coefficient γ_{i} of an added component i for the simple model of *symmetric regular solution* in a binary real mixed solid system A–B, the excess enthalpy of an added component $i = B$ is

$$\Delta h_{Bm}^{\mathrm{ex}}(x) = RT\ln \gamma_B = \Omega(1 - x_B)^2 = \Omega x_A^2 \tag{3.74}$$

where Ω is the interaction parameter comparable with energy of interaction ω [see eqs. (3.40)–(3.43)]. Inserting eq. (3.74) into eq. (3.73) at a given T, the equilibrium segregation coefficient becomes

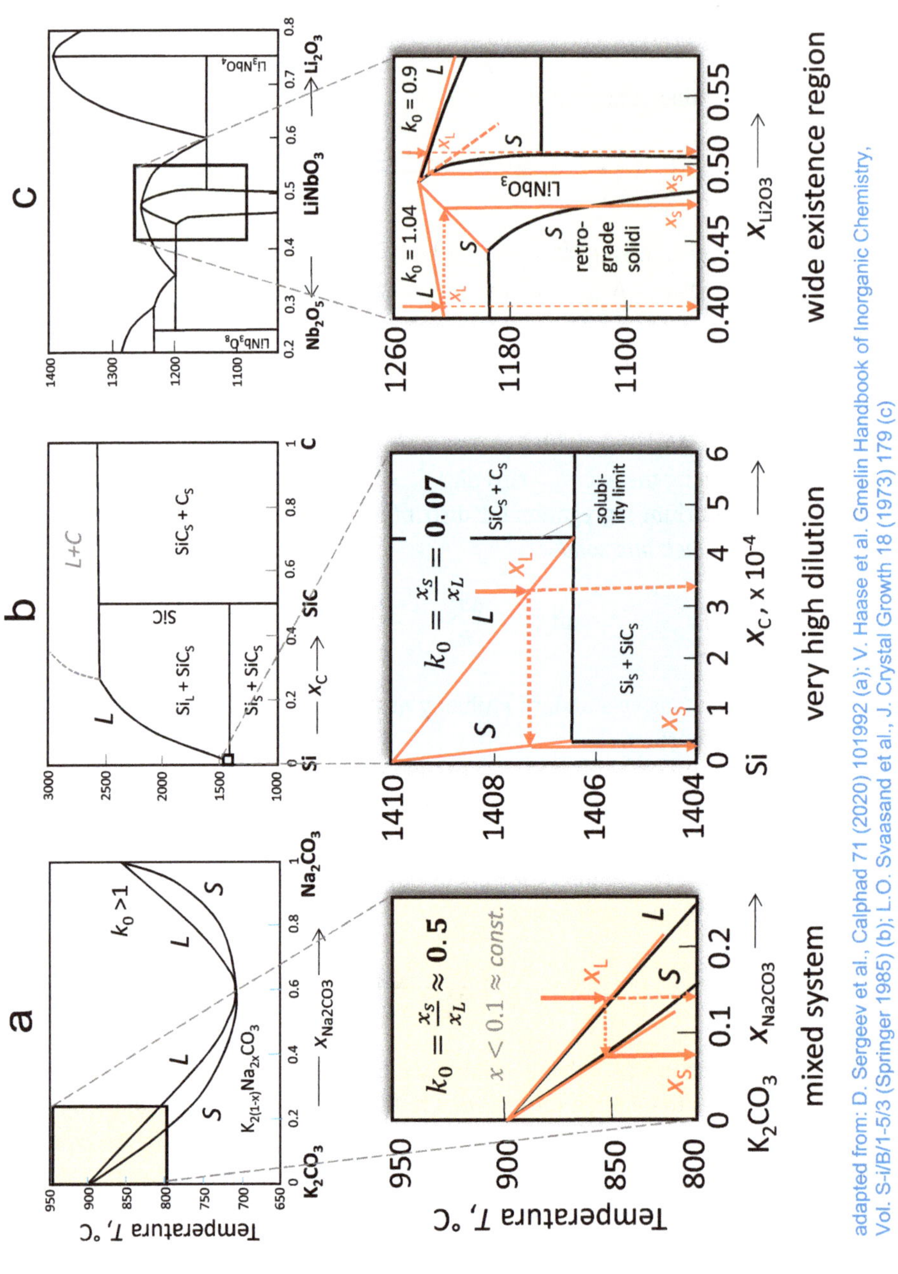

Fig. 3.27: *T*–*x* projections of selected binary systems, illustrating the segregation effect between liquidus *L* and solidus *S*, approximated by tangents; (*with permission of Elsevier (a, c) and Springer Nature (b)*).

$$\ln k_0 = \frac{\Delta h_B^0}{R}\left(\frac{1}{T} - \frac{1}{T_{Bm}}\right) - \frac{\Omega_S(1 - x_{BS})^2}{RT} \tag{3.75}$$

where Ω_s is the interaction parameter in the crystalline phase according to eq. (3.43). More detailed treatments of the interaction parameter and the related activity coefficients are given in *Spec box 3.6*. Today, there exists an enormous number of publications and textbooks dealing with experimental and theoretical (numerical) determination of the segregation coefficient in diverse melt growth material systems that are highly recommended for knowledge deepening. An approach to theoretically estimate k_0 based on the models of *Weiser (1958)* and *Chernov (1980, 1984)* is given in the *Spec box 3.9*.

As shown in Fig. 3.27a–c, mostly the constancy of k_0 is justified for low concentrations of only the additives. In reality, however, due to the redistribution during the running process of crystallization, the liquid and solid compositions successively follow the courses of the liquidus and solidus, which increasingly deviate from straight tangents with increasing concentration. As a result, the k_0 value is changing too. In Section 3.2.4.1, we presented the equations of the whole liquidus and solidus curves of an ideal mixed binary systems [see eqs. (3.56) and (3.57)]. For the first approximation, at successively selected T_{eq} values in the region between T_{Am} and T_{Bm}, each related x_{Bs}/x_{Bl} ratio and, thus the corresponding $k_0 = f(T,x)$ could be determined and then recorded into a diagram, as demonstrated by the insertion in Fig. 3.11, for example.

Spec box 3.9: Theoretical approaches to the equilibrium segregation coefficient

In the following, the theory is particularly applied to segregation in crystals of covalent bonding character. According to eq. (3.1), the chemical potential of a component *i* within a regularly mixing melt and solid can be expressed as

$$\mu_i = u_i - Ts_i + p\Omega_{Vi} \tag{B3.9-1}$$

where u_i is the specific particle energy within the phases, T the absolute temperature, s_i the specific particle entropy, p the system pressure (let us take 0.1 MPa), and Ω_{Vi} the volume of the added atom sort ≈ 10^{-23} cm^3. Because of the very low value of $p\Omega_{Vi} \approx 10^{-24}$ J (or rather in specific form by multiplying with Avogadro's constant ≈ 0.1 J/mol) this energetic part can be omitted. The specific particle energy consists of internal (potential), configurational, and deformation terms

$$u_i = u_i^{\text{int}} + u_i^{\text{conf}} + u_i^{\text{def}} \tag{B3.9-2}$$

and the specific particle entropy consists of the internal (electron state of the additive) and configurational parts as

$$s_i = s_i^{\text{int}} + s_i^{\text{conf}} \tag{B3.9-3}$$

First, the configurational energy within the crystal is determined. This term depends on the statistical distribution probability of the added atoms *B* in crystal *A*, given by the mole fraction $x_B = N_B/(N_A + N_B)$, where N_A and N_B are the total number of matrix atoms and added atoms within the matrix, respectively. As is well known from related textbooks on thermodynamics of mixed systems, the total potential energy of all bonds in a crystal of matrix *A* mixed by an additive *B* on the lattice sites yields

$$U_A^{\text{conf}} = \frac{z}{2}\left(2\omega_{AB}\frac{N_A N_B}{N_A+N_B} + \omega_{AA}\frac{N_A^2}{N_A+N_B} + \omega_{BB}\frac{N_B^2}{N_A+N_B}\right) \tag{B3.9-4}$$

where z is the number of nearest neighbors, ω_{AA}, ω_{BB}, and ω_{AB} are the bond strength-related interchange energies between similar A–A, B–B, and different atoms A–B, respectively. Thus, the change of the configurational energy by adding of one atom B into matrix A is

$$u_{B\to A}^{\text{conf}} = \left.\frac{\partial U_{sA}^{\text{conf}}}{\partial N_B}\right|_{N_A=\text{const}} \approx \frac{z\omega_{BB}}{2} + (1-x_B)^2\Omega \tag{B3.9-5}$$

with Ω the interaction parameter with the surrounding atoms, which we introduced already by eqs. (3.40)–(3.42) via the model of quasi-chemical equilibrium. The exchange (mixing) energy within the crystal with regard to one bond in intensive form is

$$\Omega = zN_{\text{Av}}\left[\omega_{AB} - \frac{1}{2}(\omega_{AA}+\omega_{BB})\right] \ [\text{J/mol}] \tag{B3.9-6}$$

(see also Spec box 3.6-ii), wherein ω_{AB} can be approximated by the empirical rule of *Allen* (1961) whereby the conventional bond energies E^b are additive:

$$E_{AB}^b \approx \frac{1}{2}\left(E_{AA}^b + E_{BB}^b\right) \tag{B3.9-7}$$

Then, the interaction energies ω_{AA} and ω_{BB} are expressed via the bond strengths between similar atoms, correlating with the sublimation enthalpies ΔH_A and ΔH_B, which are listed in tables. After carrying out this replacement and setting eq. (B3.9-7) into (B3.9-6), in consideration that the bond energy between attracting atoms is negative, it becomes

$$\Omega = \frac{(\Delta H_A - \Delta H_B)^2}{\Delta H_A + \Delta H_B} \tag{B3.9-8}$$

For instance, the sublimation enthalpy of silicon is 383 kJ/mol and of lead is 177 kJ/mol. Thus, according to eq. (B3.9-8), $\Omega \approx 76$ kJ/mol. Inserting this value into eq. (B3.9-5) with $z = 4$ due to tetrahedral coordination and assuming x_{Pb} to be very small, the configurational energy of the added element Pb in Si is $u_{\text{Pb}\to\text{Si}}^{\text{conf}} \approx 430$ kJ/mol. As can be seen from eq. (B3.9-8), the model fails when $\Delta H_A \approx \Delta H_B$, where of one can only conclude that $\Omega \ll \Delta H_A,\ \Delta H_B$.

Next is to calculate the deformation energy u_B^{def}. When a host atom A is replaced by an impurity atom B, whose normal tetrahedral radius r_B differs from that of the host atom r_A by the difference Δr (e.g., $r_B > r_A$), the four nearest neighbors move out radially by an amount Δr_1, which is less than Δr because of the compression of the impurity–host atom bonds. The displacement of the four nearest neighbors of the first shell will push the atoms of the second shell outward which, in turn, deforms the next surrounding and so forth. However, due to the rapid decrease of the deformation sphere and assuming small concentration of B without mutual influence, the first shell situation can be approximated by a continuous medium of macroscopic elastic properties, sufficiently well described by Hooke's law in the classical elasticity theory:

$$u_B^{\text{def}} = 8\pi N_{\text{Av}} G r_A \Delta r^2 \ [\text{J/mol}] \tag{B3.9-9}$$

where G is the shear modulus of elasticity of the host crystal. Again, for intensivation, the right term is multiplied by the Avogadro's constant N_{Av}. Proceeding with the same example of silicon ($G_{\text{Si}} = 80$ GPa $= 80$ kJ/cm^3, $r_{\text{Si}} = 2.35 \times 10^{-8}$ cm) doped with lead ($r_{\text{Pb}} = 2.68 \times 10^{-8}$ cm), the deformation energy is $\approx$308.6 kJ/mol.

Now, the configurational entropy is determined. The probabilities of occupation of a matrix A lattice sites by atoms B is $s = R \ \ln \ [(N_A + N_B)! \ N_A! \ N_B!]$ with $R = k \ N_{Av}$, the universal gas constant. Using Stirling's approximation ($\ln x! \approx x \ln x$), the total configurational entropy is

$$s^{conf} = -RN_A \ln \frac{N_A}{N_A + N_B} - kN_B \ln \frac{N_B}{N_A + N_B} \tag{B3.9-10}$$

The change of this value by adding one atom B into matrix A is

$$s^{conf}_{B \to A} = \left. \frac{\partial s^{conf}}{\partial N_B} \right|_{N_A = const} = -R \ \ln x_B \ [\mathrm{J/molK}] \tag{B3.9-11}$$

By the insertion of eqs. (B3.9-5), (B3.9-9), and (B3.9-11) into eq. (B3.9-1), the chemical potential of the additive B within crystal A (i.e., in the solid S) is

$$\mu_{BS} = \mu^0_{BS} + RT \ lnx_{BS} + (1 - x_{BS})^2 \Omega_S + 8\pi N_{Av} G r_A \Delta r^2 \tag{B3.9-12}$$

whereas μ^0_{BS} is the standard potential of the added element B at given temperature T consisting of

$$\mu^0_{BS} = u^{int}_{BS} + \frac{z\omega_{BB}}{2} - Ts^{int}_{BS} \tag{B3.9-13}$$

The chemical potential of the liquid phase L consists of analogous terms. However, the deformation energy can be omitted so that

$$\mu_{BL} = \mu^0_{BL} + RT \ln x_{BL} + (1 - x_{BL})^2 \Omega_L \tag{B3.9-14}$$

where Ω_L is the interaction parameter in the melt consisting of the coordination number z_L in the melt (~6) and mixing energy in the melt, analogous to eq. (B3.9-6). Usually, this value is in most melts relatively small, yielding only few kJ/mol, e.g., in molten silicon, 5–25 kJ/mol.

In the case of relatively small impurity/dopant concentration, it is $(1 - x_{BS,L}) \approx 1$. Further, according to eq. (3.53), the difference of the standard potentials is $\mu^0_{BL} - \mu^0_{BS} = \Delta h_B - T\Delta s_B$. After equating eqs. (B3.9-13) and (B3.9-11), it becomes

$$\ln \frac{x_{BS}}{x_{BL}} = \ln k_0 = \frac{(\Delta h^0_B - T\Delta s^0_B) + \Omega_L - \Omega_S - u^{def}_B}{RT} \tag{B3.9-15}$$

Setting $\Delta s_{B0} = \Delta h_{B0} / T_{Bm}$, with T_{Bm} the melt temperature of the pure added element B, the segregation coefficient is given by

$$k_o = \exp \left[\frac{\Delta h^0_B}{R} \left(\frac{1}{T} - \frac{1}{T_{Bm}} \right) + \frac{\Omega_L - \Omega_S - u^{def}_B}{RT} \right] \tag{B3.9-16}$$

In the literature, one can find diverse theoretically calculated k_0 values and their comparison with experimentally determined ones. Here are some examples: Sn in Si $\rightarrow k_{calc} = 0.8 \times 10^{-2}$, $k_{exp} = 2 \times 10^{-2}$; As in Si $\rightarrow k_{calc} = 1 \times 10^{-2}$, $k_{exp} = 9 \times 10^{-2}$; Pb in Ge $\rightarrow k_{calc} = 1 \times 10^{-4}$, $k_{exp} = 4 \ \times 10^{-4}$.

Let us remain at the small impurity/dopant concentration levels. Then, the mathematical approach proves to be considerably simpler. From T–x phase projections, it follows that the additive increases or decreases the melting point of the matrix element. The equilibrium of chemical potentials can then written as

$$\mu_{AS} + RT \ln(1 - x_{BS}) = \mu_{AL} + RT \ln(1 - x_{BL}) \tag{B3.9-17}$$

and at very small concentration of B, one can approximate

$$\ln\frac{(1-x_{BS})}{(1-x_{BL})}\approx(x_{BL}-x_{BS}) \qquad \text{(B3.9-18)}$$

from which follows at use of $k_o = x_{BS}/x_{BL}$ that $(x_{BL}-k_o x_{BL}) = x_{BL}(1-k_o)$. Inserting this relation into eqs. (B3.9-18) and (B3.9-17) and setting $\mu_{AL}-\mu_{AS}=\Delta h_A - T\Delta s_A = \Delta h_A(T_{mA}-T)/T_{mA}$, it becomes

$$k_0 = 1-\frac{\Delta h_A(T_{mA}-T)}{RTT_{mA}x_{BL}} \qquad \text{(B3.9-18)}$$

Due to the validity of eq. (B3.9-18) only for $x_B \ll x_A$, one can equate $T \approx T_{mA}$, leading to the final quite practicable form, deduced by *Hayes* and *Chipman* (1939) as

$$k_0 = 1-\frac{\Delta h_A \Delta T}{RT_{mA}^2 x_{BL}} \qquad \text{(B3.9-17)}$$

This formula proves to be very helpful for the crystal grower to estimate the expected distribution along the as-grown crystal and also the effectiveness of material purification.

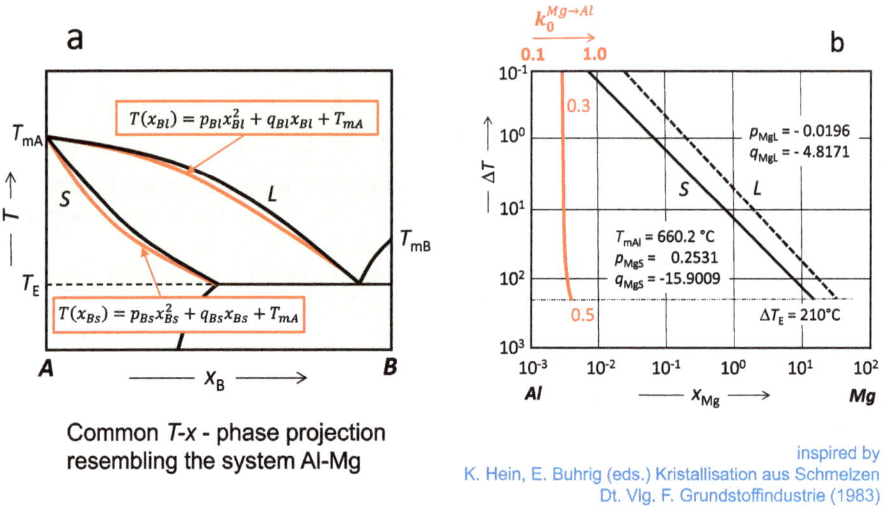

Fig. 3.28: Sketched regression functions of liquidus and solidus in a characteristic binary system (a) and in the double-log coordinated *T*–*x* projection of the system Al–Mg with determined k_0 (b).

An exact determination of k_0 on the basis of known phase projections proves to be the description of the real solidus and liquidus courses by using the *regression functions*. As is demonstrated by Fig. 3.28a, it can be assumed that in most phase diagrams, the trajectories of solidus and liquidus can be expressed by polynomials. For instance, a polynomial of second order describes the temperature dependence of the *solidus* as a function of the mole fraction of the added component *B* as

$$T(x_{BS}) = p_{BS}x_{BS}^2 + q_{BS}x_{BS} + T_{mA} \tag{3.76}$$

and of the *liquidus* as

$$T(x_{BL}) = p_{BL}x_{BL}^2 + q_{BL}x_{BL} + T_{mA} \tag{3.77}$$

where $p_{BS,L}$ and $q_{BS,L}$ are the related regression coefficients and T_{mA} is the melting point of the crystalline matrix (note, due to the corresponding adjustment, the temperature T and mole fraction in eqs. (3.76) and (3.77) have the dimensions °C and mol%, respectively). The validity range of regression is between T_{mA} and T_E (eutectic temperature). Thus, the deviation of the given temperature in this region $T(x_{BS,L})$ from the melt temperature T_{mA} is $\Delta T_{S,L} = T_{mA} - T(x_{BS,L}) = -p_{BS,L}x_{BS,L^2} - q_{BS,L}x_{BS,L}$. Because the equilibrium segregation coefficient $k_0 = x_{BS}/x_{BL}$ always represents the isothermal concentration ratio so that $\Delta T_S = \Delta T_L$, one becomes $p_{BS}x_{BS^2} + q_{BS}\,x_{BS} = p_{BL}x_{BL^2} + q_{BL}\,x_{BL}$, from which the concentration dependence of k_0 can be derived as

$$k_0(x_B) = \frac{p_{BL}x_{BL} + q_{BL}}{p_{BS}x_{BS} + q_{BS}} \tag{3.78}$$

where the regression coefficients should be taken from the experimentally determined T–x projections according to the method of the least square error. By an analogue procedure, the temperature dependence of the segregation coefficient can be obtained as

$$k_0(T) = \frac{p_{BL}\left(\pm\sqrt{q_{BS}^2 - 4p_{BS}\Delta T}\right) - q_{BS}}{p_{BS}\left(\pm\sqrt{q_{BL}^2 - 4p_{BL}\Delta T}\right) - q_{BL}} \tag{3.79}$$

where $\Delta T = T_{mA} - T(x_{BS,L})$. For instance, in the system Al-Si within the temperature region 660 °C (T_{mA}) – 577 °C, the regression coefficients are $p_{SiL} = -0.0371$, $q_{SiL} = -6.2958$, $p_{SiS} = 5.6065$, and $q_{SiS} = -77.1694$ so that at $T(x_{BS,L}) = 600$ °C the segregation coefficient becomes $k_0 \approx 0.1$, being in good agreement with experimental data.

Because of the difficult readout of small concentrations in a common T–x phase projection with linear coordinates (Fig. 3.28a), it is more favorable to present the regression functions in log-log coordinates as is demonstrated for the system Al–Mg in Fig. 3.28b. The equilibrium segregation coefficient of Mg in Al, determined in this way, is added as red curve in logarithmic scale. Its functional dependence on the melting point reduction $k_0(\Delta T)$ shows nearly constant value of 0.3 up to $\Delta T = 10^2$ but then increases up to 0.5 at $(T_{mAl} - T_E) = \Delta T_E = 210$ °C where T_E is the eutectic temperature = 450 °C (quite comparable with Fig. 3.2a).

Such a k_0 determination via regression functions may be very helpful for the growth of mixed crystals within a wide range of mole fraction, like binary $Cu_{1-x}Ni_x$, $Ge_{1-x}Si_x$, and pseudo-binary systems $In_{1-x}Ga_xP$, $Cd_{1-x}Zn_xTe$, $Ba_{1-x}K_xBiO_3$, for example. A significant practical challenge proves to be the management of homogeneous x distribution within the growing crystal as far as possible (we will return to this important task in the lecture part III). On the other hand, crystal growth also means obtaining monocrystalline objects of certain physical properties requiring highest chemical purity. Therefore, the minimization of residual impurities and homogeneous distribution of dopants of relatively low concentrations (10^{12}–10^{17} cm^{-3}) plays a similarly important role. At such highly *diluted solution*, named *Henry's range*, the interaction between the impurity and dopant atoms or molecules can be neglected and the above mentioned approximation of the liquidus and solidus by two tangents meeting in the melting point of the matrix component A (see Fig. 3.27a) is quite justified. Then, k_0 can be treated as constant and determined by

$$k_0 = \frac{x_{BS}}{x_{BL}} = \frac{n_{BS}}{n_{AS} + n_{BS}} \Big/ \frac{n_{BL}}{n_{AL} + n_{BL}} = \frac{n_{BS}}{n_{BL}} = \frac{c_{BS}}{c_{BL}} = \frac{w_{BS}}{w_{BL}} \approx \text{const} \tag{3.80}$$

where c_B is the concentration, n_B is the component quantity, and w_B is the mass fraction of the regarded impurity or doping element B [see eqs. (3.13)–(3.15)].

At the growth of compound crystals, one is concerned with the problem of segregation of the excess partner component A or B between the liquidus and solidus of the existence region $(A_{\nu_A \pm \delta x_A} B_{\nu_B \mp \delta x_B})_S$ [see eq. (3.60) and Figs. 3.13, 3.14, and 3.27c]. This aspect becomes more complicated when the congruent melting point is deviated from stoichiometry and the solidus course above the eutectic point is retrograde, as was already sketched in Fig. 3.13. Incidentally, the solubility curves of impurities/dopants in any crystalline material show always such retrograde behavior too. Fig. 3.29a represents the magnified existence region of CdTe, combined with the Te-rich side of the T–x phase projection. The larger the deviation from stoichiometry, the more $k_{0\mathrm{Te}}$ deviates from unity. Whereas near the CMP its value yields about 0.3, at the maximum width of the existence region it is reduced to 10^{-3} – a marked divergence. Fig. 3.29b shows the solubility curves (solidus) of selected impurity/doping elements in silicon. Also, in these cases, the related equilibrium segregation coefficients show a distinct temperature and concentration dependence. As an example, for the calculation of the solubility curve of Cu, the regression coefficients are added. The value of the limiting distribution coefficient at $x_{\mathrm{Cu}} \to 0$, e.g., at very small concentrations is $k_{0\,\mathrm{limCu}} = 2 \times 10^{-4}$. It plays an important role with regard to ultrahigh material purification.

In principle, the exact determination of the $k_0(T)$ function within the whole range between melting point of matrix A and eutectic temperature ΔT_E would need a careful regression analysis (there are related commercial codes, e.g., packet Mathcad). For ideal diluted solutions, the simplified equation of *Romanenko* (1960) can be applied in the following form:

$$k_0 = \frac{k_{0\,\mathrm{lim}B}}{(1 - x_{BL})^{a}} \tag{3.81}$$

where the exponent a is determined by the formula

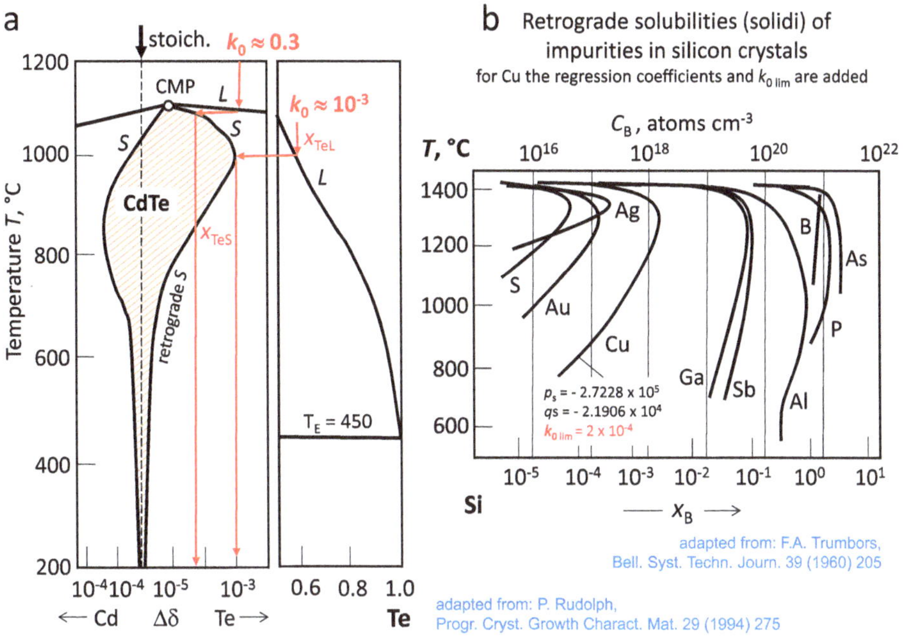

Fig. 3.29: Segregation phenomenon in systems with retrograde solidus lines, e.g., CdTe (a) and impurities in silicon (b) (*with permission of Elsevier (a); John Wiley and Sons (b)*).

$$a = \left(\frac{1}{\Delta S_{mA}} (R \ln k_{0\,\lim B} + \Delta S_{mB}) \right) \tag{3.82}$$

where R is the universal gas constant and $\Delta S_{mA,B}$ are the melting entropies of A and B components, respectively. The exponent a for each pair of A–B can have a positive as well as negative value. Negative exponents reflect a retrograde solubility as in the case of Cu (and another impurities) in Si, shown in Fig. 3.29b.

Usually, at standard melt growth production of compound materials, the samples weights show only very small deviations from stoichiometry or from congruent melting point, so that one can certainly assume that only a very small excess content of one of the components takes place. As a result, the solidus and liquidus lines are positioned still close to each other and can be linearly approximated, resulting in a constant equilibrium segregation coefficient, mostly very close to unity. For instance, Fig. 3.27c shows the segregation coefficient of Li_2O_3 in $LiNbO_3$ being constant = 0.9, up to ~1% deviation from stoichiometry. Nearly the same situation exists in the growth of GaAs crystals provided from melt, with small As excess.

A further aspect that must be perhaps considered is the possible mutual interaction between doping elements when more than one of them is added in relatively high concentrations. Besides the component of matrix A, the additives $B, C, \ldots, N_i$ are presented within the melt and solid. Now, it can happen that an attraction between the dopants leads to the formation of associates with chemical bonds. Such species affect the value of the segregation coefficient, especially when the associates as a whole are incorporated into the growing crystal. Provided that one knows the form of associates and their integers of stoichiometry [see. eq. (3.58)], the concentrations in the ratio of $k_0 = c_S/c_L$ must be replaced by the sum of the associates within the liquid and solid as $c_{N_i S,L} = \sum_C^{N_i} \left[(A_{\nu_A} B_{\nu_B})_{S,L} \ldots (A_{\nu_A} N_{i\nu_B})_{S,L} \right]$. Such quite a rare situation was treated in detail by *Vigdorovich* (1969) in his book on purification of metals and semiconductors given in the references at the end of this lecture part. As was pointed out by *Chernov (1984)*, the behavior of doped material in compound semiconductors is particularly complicated by their interaction with intrinsic point defects X_i (interstitial and vacancy), which are near the melting point, isolated and usually electrically charged. As a result, the dopant solubility can be markedly influenced by complex formations with the charged intrinsic point defects such as $[B^{1+} - X^{3-}]^{2-}$, for example, that would lead to differing measured values of the segregation coefficient, depending on whether it was determined by mass or electrical analysis.

Generally, in the melt growth of contaminated and intentionally doped compound crystals, each crystal grower must consider the acting mass-related segregation effect very carefully because at a propagating melt–solid interface, the enrichment of an excess component or dopant at the crystallization front and its rejection back to the melt within a diffusion boundary layer may cause an undesirable morphological instability (see lecture parts III).

3.2.6.2 Segregation at solution–solid phase transition

In solution growth, quasi three components – the *solvent*, the *solid* (crystallizing material), and the *solute* (matrix material components, impurities and dopants in solution) – are presented. Of course, it makes sense to apply only such solvent which shows high solubility in the fluid (solution) but being as much as possible, insoluble in the crystalline matrix. This proves to be one of the challenges before the growth is started, to find out the best solvent. Principally, considering the whole very wide branch of solution growth, the incorporation of impurities or dopants into the growing crystals is a complex combined effect of several factors, like the chemical composition of solution, relative solubilities of host and impurity/dopant phases, interactions between host and impurity atoms/molecules, relative dimensions of substituting (impurity/dopant) and substituted (host phase) ions, the similarity in crystallographic structure of the two phases, and crystallization conditions such as growth temperature, supersaturation used for the growth, and concentration of an additive. All these facts determine the actual segregation coefficient of solution–solid transitions. However, the treatment of all these miscellaneous factors here would be too much. Reference is therefore made to the large respective literature.

Relatively well-comprehensible is the area of growth from *melt–solutions*, often applied for metals and semiconductors. We will limit ourselves to this branch (although the presented approach to the segregation coefficient is also applicable to another simple solution growth cases). In addition to the matrix element $A_{S,L}$ (in mole fraction $x_{AS,L}$) to be crystallized and the added solute (impurity and dopant), $B_{S,L}$ ($x_{BS,L}$), the liquid phase also consists of the solvent $C_L(x_{CL})$, which is usually not solved in the solid ($x_{CS} \approx 0$, $k_0^C \ll 1$). Actually, things are then quite easy. Whereas in the crystal, the impurity/dopant solubility B concerns only its relation to the matrix A, in the liquid solution, its content is related to the sum of all dissolved components, i.e., B and A, both solved in C. The equilibrium segregation coefficient K_0 of a given additive (solute) B is then

$$K_0 = \frac{x_{BS}}{\overline{X_{BL}}} \tag{3.83}$$

where $\overline{X_{BL}}$ is the concentration ratio of the solute B_L to the sum of the solved elements A_L and B_L, whereby at very low concentration of B_L the concentration of the predominant solved component A_L in solution C_L can be expressed by the mole fraction $x_{AL} \approx 1 - x_{CL}$ so that

$$\overline{X_{BL}} = \frac{x_{BL}}{(x_{AL} + x_{BL})} = \frac{x_{BL}}{[(1 - x_{CL}) + x_{BL}]} \tag{3.84}$$

Due to $x_{BL} \ll x_{AL}$, eq. (3.84) can be approximated as

$$\overline{X_{BL}} = \frac{x_{BL}}{x_{AL}} = \frac{x_{BL}}{(1 - x_{CL})} \tag{3.85}$$

with x_{AL} the mole fraction of the matrix component A_L solved in the solvent C_L. Substitution of (3.85) into (3.83) results in the *equilibrium distribution coefficient at solution–solid transition*

$$K_0 = \frac{x_{BS}}{x_{BL}} x_{AL} = \frac{x_{BS}}{x_{BL}} (1 - x_{CL}) = k_0 (1 - x_{CL}) \tag{3.86}$$

where k_0 is the equilibrium segregation coefficient of the given impurity or dopant element near the melting point, i.e., at melt–solid transition [eq. (3.80)], assuming that the concentrations $[B] \ll [A]$, $[C]$ (in the literature, often the symbol $\boldsymbol{D}_o$ is used instead of K_0).

From eq. (3.86), it follows that in growth from solutions, the segregation coefficient of a given added element (impurity and dopant) K_0 is reduced in comparison to the growth from melt when $k_0 < 1$. In other words, the effect of impurity purification is enhanced in growth from solution due to their increased solubility and, thus, retention in the liquid solution. In fact, significant purification effect in growth from melt–solution has been described in the literature. This is also traceable when looking at the ternary-phase diagrams in Fig. 3.25 though close to the B corner.

In some cases, the solubility of the solvent in the crystalline phase cannot be neglected. Then, eq. (3.83) must be modified by replacing the denominator x_{BS} with $\overline{X_{BS}}$, the concentration ratio of the solute B_S to the sum of solutes ($A_S + B_S$). In fact, such situation takes place when compound crystals AB are growing from melt–solutions in which one constituent A or B acts as solvent. Such growth principle is commonly used by the LPE or THM techniques to many compound and mixed crystals, like AlSb from Al, (Ga,Al)As from Ga, GaN from Ga, (Cd,Zn)Te from Te, or even oxides by top-seeded solution growth (TSSG) like (Ba,Sr)TiO_3 from TiO_2, $LiBaF_3$ from LiF, a.s.o. In order to prevent confusion with the solvent B in such cases, the symbol for the impurity/dopant should be replaced by a differing character $\mathfrak{z}$, for example. Due to the width of the compound existence region $\pm\Delta\delta$ [see Section 3.2.4.2.1; eq. (3.60)], in addition to the impurity/dopant $\mathfrak{z}$, a certain excess of the solvent A or B is also solved in the solid phase AB (see, e.g., Fig. 3.29a). It is therefore possible that an excess of solvent B is realized by vacancies in the sublattice A, as often observed in reality. As a result, the vacancies are filled with impurity/dopant atoms $\mathfrak{z}$, which means that their incorporated content is dependent on the degree of deviation from stoichiometry. In other

Tab. 3.3: Comparison of the experimentally obtained equilibrium segregation coefficients at growth from melt-solution K_0 with the segregation coefficient near the melting point of the matrix crystal k_0 for selected materials, where the K_0 value is reduced compared to $k_0 < 1$ (*with permission of Springer Nature, Elsevier (Science Direct), and Wiley*).

Crystal (matrix A)	Additive $\mathfrak{z}$ in matrix A (impurity/dopant)	k_0 of $\mathfrak{z}$ at T_m (melt growth)	Solution growth method	Solvent B	K_0	Relation K_0/k_0	References
Si	In	**4×10^{-4}** **2.5×10^{-4}**	THM 950–1,300 °C	In	**1×10^{-6}** **1×10^{-6}**	0.02 0.04	W. Scott, R.J. Hager, J. Electron. Mat. 8 (1979) 581
GaAs	Sn	**3×10^{-2}**	LPE 700–800 °C	Ga	**1.2×10^{-4}**	0.04	J. Vilm, J.P. Garrett, Solid-State Electronics 15 (1972) 443.
ZnSe	Ga Cu	**0.3** **0.11**	LPE 840–950 °C	Sn	**0.1** **0.04**	0.33 0.36	T.F.McGee et al., J. Crystal Growth 59 (1982) 649
CdTe	In Cl Ag Mg	**0.43** **0.36** **0.2** **3.22**	VB from Te-rich melt–solution 880 °C	Te	**0.06** **0.005** **0.01** **<1.5**	0.14 0.01 0.05 0.46	K. Zanio, in: R. Willardson, A. Beer (eds.) Semicond. and Se-mimetals 13 (Acad. Press, N.Y. 1978).
YAG	Nd ($x = 0.05$) Nd ($x = 0.32$)	**0.19**	Dipping LPE 925–940 °C	B_2O_3–PbF_2–PbO	**0.025** **0.180**	0.13 0.90	M. Sturge et al., Mat. Res. Bull. 7 (1972) 989

The k_0 values in the matrix materials near melting point T_m are taken from: V.M. Glazov, V.S. Semskov, Fizikokhimich. osnovy legirovanija poluprov (Nauka 1967); O. Madelung, Physics of III–V Compounds (J. Wiley 1964); H. Hartmann et al., in: E. Kaldis (ed.) Current Topics in Mat. Sci. 9 (North-Holland 1982); L. Kuchar et al., J. Crystal Growth 161 (1996) 94; M. Sturge et al., Mat. Res. Bull. 7 (1972) 989.

words, the segregation coefficient of $Ƶ$ is a function of the deviation from stoichiometry, the degree of which is determined by the growth temperature (see Figs. 3.13 and 3.19a). In fact, years ago, the author and his team detected, by PL, an increasing concentration of silver atoms on the sides of the Cd-sublattice in CdTe crystals with increasing deviation from stoichiometry toward Te-rich site.

Figure 3.30 shows a practical example of the reduction of the axial dopant distribution curves for important donors In, Cl, and Sn in a CdTe crystal in the growth with a melt–solution zone of the solvent tellurium by THM. In comparison with the single-pass curves for the melt zone growth, inserted in the graphic at the top right (well-known as Pfann distribution curve), the distribution coefficient k_0 was replaced by the equilibrium distribution coefficient for solution–solid transition K_0 according to eq. (3.86).

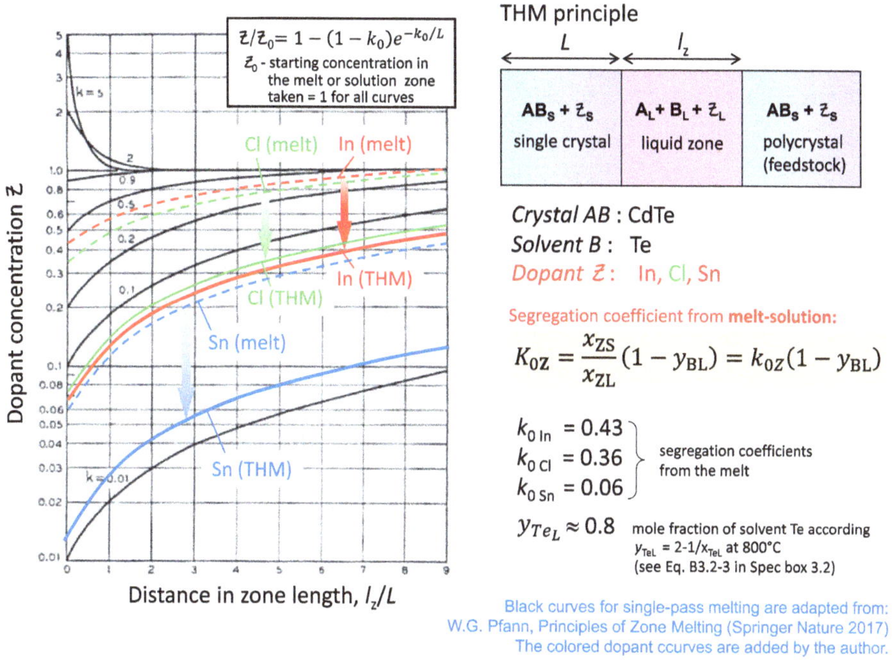

Fig. 3.30: Reduction of axial dopant distribution in a CdTe crystal after single pass of a Te-rich melt–solution zone compared to the melt zone (*with permission of Springer Nature*).

Since the solvent Te is a component of the *AB* compound crystal CdTe, it becomes here the notation *B* and the dopants $Ƶ$. Selecting the melt–solution zone temperature 800 °C, the mole fraction of Te in the system Cd–Te is $x_{Te} = 0.83$. However, due to the solvent role of Te versus the compound CdTe, its mole fraction calculates within half of the system *AB*(CdTe)–*B*(Te) according to eq. (B3.2-3) in *Spec box 3.2* to be $y_{Te_L} = 2 - 1/x_{Te_L} \approx 0.8$ and, thus, the K_0 values of the dopants $Ƶ$ becomes, according to eq. (3.86), $K_0 = k_0(1 - y_{Te_L}) \approx 0.2\ k_0$.

In fact, the obvious enhanced purification effect at the growth from melt–solution proves to be favorable for impurity cleaning but adverse for the axial distribution of dopants, showing an enhanced inhomogeneity even at very small K_0 values, like in the case of Sn (see Fig. 3.30).

Note, when the additive B is a mixed crystal component in the matrix A as $A_{1-x}B_x$, usually its mole fraction share x_{B_L} is no longer negligible as was assumed by eq. (3.84). Therefore, eq. (3.86) must be extended by the addition of the quotient x_{B_S}/k_0. This is of significance for the popular growth of $Cd_{1-x}Zn_xTe$ crystals from tellurium-rich melt–solution by THM, taking into account that k_0 (Zn) > 1.

3.2.6.3 Segregation at vapor–solid phase transition

Usually, at the transition from the vapor to the solid phase, the effect of segregation is much more pronounced. The description of the concentration difference of a given added element B (impurity and dopant) in a matrix A via an equilibrium coefficient of segregation (or distribution) between solid (S) and vapor (V) phases is

$$k_0^{\mathrm{VS}} = \frac{x_{BS}}{x_{BV}} \tag{3.87}$$

with $x_{BS} = n_{BS}/(n_{BS} + n_{AS})$ and $x_{BV} = n_{BV}/(n_{BV} + n_{AV})$ the mole fractions (given here in component quantities $n_{iS,V}$) in the solid and vapor phases, respectively. However, whereas the indication of the component quantity (or concentration) in the crystalline phase is relatively simple, its determination within the gas phase proves to be somewhat more complicated. First of all, one has to consider the incommensurable much lower density of each vapor phase, compared to the condensed one ($\rho_V \ll \rho_{S,L}$). Then, only within a closed system, where the growing crystal and vapor are enclosed in an ampoule or tight container, an equilibrium situation can be assumed. In comparison, in open gas transport systems, the quantities (densities) of the matrix element to be crystallized and the intentional as well as the unintentional additives quasi predetermined in transport agents are mostly not in equilibrium (supersaturated) with the crystal state. Therefore, the strong thermodynamic x_{BS}/x_{BV} equilibrium will occur only within the interface region and correlates with the kinetic processes such as surface diffusion, adsorption-desorption, and atomistic interface structure (see lecture part II).

However, a phenomenological treatment is quite possible. In the vapor–solid phase transition, it makes sense to correlate the segregation (distribution) with the *partial vapor pressures* of the constituents within the gas phase. Usually, they differ significantly. For instance, whereas the partial pressure of solid silicon at 1000 °C is $p_{\mathrm{Si}} = 10^{-8}$ at ($\approx 10^{-4}$ Pa), at the same temperature, Cu, Al, and Fe evaporate with $p_{\mathrm{Cu}} = 10^{-4}$ at (≈ 10 Pa), $p_{\mathrm{Al}} = 10^{-5}$ at (≈ 1 Pa) and $p_{\mathrm{Fe}} = 10^{-6}$ at ($\approx 10^{-1}$ Pa), respectively. Regarding these metals as impurities in silicon, a marked *separation effect* is occurred at its sublimation. Therefore, the sublimation-condensation in an open system is used as very effective measure of element purification already for a long time. Accordingly, it is more convenient to replace in the relation of the segregation coefficient [eq. (3.87)], the mole fraction (component quantity) of the vapor phase by the relation of the *Raoult's Law* $p_B = x_{BV}\, p_{0B}$ with p_B the partial pressure of the component B in the gaseous mixture, and p_{0B} the vapor pressure of the pure component B (see Fig. 3.15). Then, it becomes

$$x_{BV} = \frac{n_{BV}}{n_{BV} + n_{AV}} = \frac{p_B}{p_{0B}} \tag{3.88}$$

and

$$k_0^{VS} = \frac{x_{BS}}{x_{BV}} = \frac{\frac{n_{BS}}{n_{BS} + n_{AS}}}{\frac{n_{BV}}{n_{BV} + n_{AV}}} = \frac{x_{BS}}{\frac{p_B}{p_{0B}}} = \frac{x_{BS} p_{0B}}{p_B} \tag{3.89}$$

The distribution coefficient at vapor–solid transition plays an important role in the treatment of epitaxial processes of ternary mixed crystal systems. Its proper control ensures the layer deposition with homogeneous composition. *Stringfellow (1980)* calculated the vapor–solid distribution coefficient for metal-organic chemical vapor phase deposition (MOCVD) of mixed III–V semiconductor systems, such as $Ga_{1-x}Al_xAs$. Using trimethylgallium (TMGa), trimethylaluminium (TMAl), and AsH_3 as transport agents and assuming the equality between the diffusion coefficients of Ga and Al in the vapor, the ratio of x_{AlAs_S} to x_{Al_v} was found to be approximately unity.

$$k_0^{VS} = \frac{x_{AlAs_S}}{x_{Al_v}} = \frac{x_{AlS}}{\left[p_0^{TMAl} / \left(p_0^{TMAl} + p_0^{TMGa}\right)\right]} \approx 1. \tag{3.90}$$

A similar k_0^V value was found for all alloys, where mixing occurs on the group III sublattice, such as $Ga_{1-x}Al_xSb$ and $In_{1-x}Ga_xAs$, for example. At normal growth temperatures, the partial pressures of Al, Ga, and In proved to be nearly zero at the growing interface with V/III ratio $\gg 1$. In comparison to that at the growth of alloys with mixing on the group V sublattice, such as $GaAs_{1-x}Sb_x$, the value of k_0^V was found to be smaller than unity, especially at lower temperatures.

3.2.6.4 Segregation at vapor–liquid-phase transition

In all crystal growth processes from melt and solutions, a further phase boundary has to be considered – the vapor–liquid interface. Also, a separation effect can occur here. This concerns the incorporation of impurities or dopants from the gas phase or their evaporation, often technically used in melt refining processes. As was already discussed in Section 3.2.4.2.3, in some vertical melt-growth arrangements, like VB, VGF, and Kyropoulus techniques, there is a missing direct contact of the melt–solid interface with the gas atmosphere. In such cases, the front of crystallization is totally covered by the melt column. Especially, for the control of stoichiometry in growing compounds by the overpressure of the volatile component, the vapor species must overcome the effect of vapor-melt separation, their transport through the melt toward the growing crystal, and the melt–solid segregation at the propagating interface (see Section 3.2.4.2.3). Additionally, one has to consider the volatility of doping and impurity elements even through the liquid-gas interface that influences their assumed starting concentration, being important for the estimation of the axial and radial segregation curves in the growing crystal. Such a system is referred to as an open or *non-conservative* (more details are given in the lecture part III).

The segregation effect at the vapor–liquid interface can be estimated as following (see also *Spec box 3.6*). When a substance B (dopant and impurity) goes into solution with a solvent of the matrix element A at temperature T (for instant at melt temperature of A), the partial vapor pressure values of both elements A and B change from $p^0_{B,A}$ to $p^{eq}_{B,A}$. If the vapor above the solution shows an ideal behavior, the ratio of the vapor pressures represents the B and A activities as

$$a_{B,A} = \frac{p^{eq}_{B,A}}{p^0_{B,A}} = \gamma_{B,A} x_{B,A} \tag{3.91}$$

where $\gamma_{B,A}$ is the activity coefficients and $x_{B,A}$ the related mole fractions. Under equilibrium conditions without T inhomogeneities along the vapor–liquid interface becomes

$$\frac{p^{eq}_B}{p^{eq}_A} = k^{VL}_B \frac{x_B}{x_A} \tag{3.92}$$

where k^{VL}_B represents the *separation (segregation) coefficient* of the added element B between vapor and liquid as

$$k^{VL}_B = \frac{p^0_B \, \gamma_B}{p^0_A \, \gamma_A} \tag{3.93}$$

If k^{VL}_B is equal to unity, the concentrations of B in liquid and vapor are equal, and no separation occurs. When $k^{VL}_B < 1$, the equilibrium concentration of B in liquid A is more than that in the gas phase. As a result, an impurity B cannot be separated by partial evaporation. In contrast, when $k^{VL}_B > 1$, the separation of an impurity B from a liquid A is possible. For instance, in molten silicon of Raoultian behavior ($\gamma_A = \gamma_{Si} \approx 1$), the activity coefficients γ_B of Al, Fe, Na, and P are 0.37, 0.014, 0.466, and 0.4522, respectively. Thus, purification from these elements by evaporation would not be possible. On the other hand, an effective fractionation turns well for the elements Zn, Bi, and B showing γ_B values of 1.471, 29.68, and 3.896, respectively. This proves to be of importance for the production of high-purity silicon feedstocks for electronics and photovoltaics, for example.

4 Surfaces, phase boundaries, and interfacial effects

4.1 Determination of the surface free energy

Crystal growth processes are interconnected with propagating phase boundaries or rather *interfaces*. In case of vapor–solid transition this is realized by the *surface* of the growing crystalline phase. At melt–solid, solution–solid, and solid–solid transition this is often termed *front of crystallization*. One of the targets of thermodynamics considers the equilibrium shape of a growing crystal surrounded by a high-purity homogeneous voluminous fluid mother phase.

Each surface of a crystal is a cross-discontinuity and has a free energy associated with it. The value of this free energy depends on the orientation and specific state of the face and on the other phase in contact (vacuum, fluid). Dangling unsaturated bonds enhance the free potential of Gibbs as sketched in Fig. 4.1. As a result various inherent processes of energy minimization are activated. For instance, the interatomic forces drive the free bonds to saturate jointly by turning into the given surface plane. As a consequence, the atoms near the surface modify their positions differing from spacing and symmetry of the bulk atoms, creating a different surface structure with enhanced *Gibbs surface potential* ΔG^{surf} (see Fig. 4.1). Such change in equilibrium positions near the surface represents a kind of relaxation, which belongs to atomically clean surfaces in vacuum and is categorized as *reconstruction*. On the other hand, at the interaction with another medium the relaxation *by adsorption* occurs. Thus, the energetic situation along the surface area is changed compared to the crystal volume, leading to the formation of a *surface energy*. According to a general definition, it is the reversible work required to create a unit area of surface. The surface energy defined per unit area is comparable to the *surface tension*. Surface tension is defined as the force parallel to the surface perpendicular to a unit length line drawn on the surface and, therefore, also defined as energy per unit area.

When the surface is in contact with another solid/crystalline phase, e.g., at heteroepitaxial thin-film arrangements, the free surface energy results from the differences in the crystallographic structure and tendencies of each solid phase to attract its own atoms or molecules and it is more correct to name it *interface energy*. According to the simple Gibbs model this energy of a very thin transitional region with mostly misfit restrain results from the difference between the internal total system energy and the energy contributions of the two contacting solid phases per interface area as sketched in Fig. 4.1. Of course, there is also an interface between a solid and liquid or vapor phase.

Therefore, in the processes of crystallization we have to consider this new surface and interface-related thermodynamic parameter. That means the potential functional-

https://doi.org/10.1515/9783111711164-004

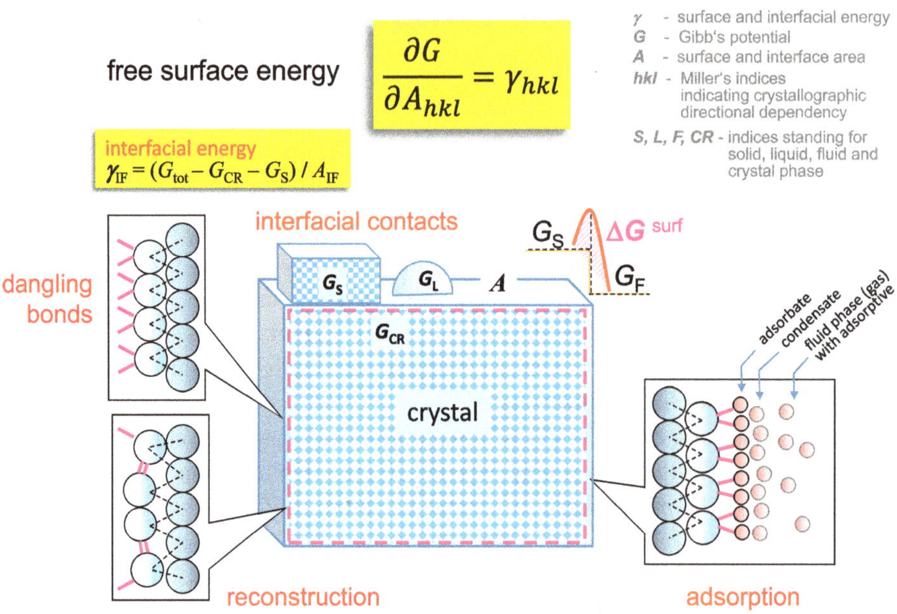

Fig. 4.1: Surface and interfacial effects on a crystal contour requiring the consideration of a new energetic parameter γ.

ity of the Gibbs free energy in eq. (3.9) must be extended by the parameter of surface area A as

$$G = f(T, p, n_i, A) \tag{4.1}$$

In compliance with eq. (3.10), the total differential of the Gibbs free energy is then

$$dG = \left.\frac{\partial G}{\partial p}\right|_{T,n_i} dp - \left.\frac{\partial G}{\partial T}\right|_{p,n_i} dT + \sum_{i=A}^{K} \left.\frac{\partial G}{\partial n_i}\right|_{p,T} dn_i + \left.\frac{\partial G}{\partial A}\right|_{p,T} dA \tag{4.2}$$

whereas the new differential quotient means that for an enlargement of the area A by the differential value dA, a work needs to be done against the surface (or interface) activity (or tension) expressed by the surface-related partial derivative

$$\left.\frac{\partial G}{\partial A}\right|_{p,T} = \gamma \tag{4.3}$$

where γ is the *surface free energy* or *surface tension* (note, in the literature γ is often replaced by the symbols σ or α).

The *surface energy* is defined as the energy per unit surface area, i.e., of the bond strength along the surface. It can be defined as the energy difference between the bulk of the material and the surface of the material. Otherwise, the *surface tension* is mostly used to define the net intermolecular force on the surface molecules of a liquid. Compared to the molecules in the center of a liquid volume those on its surface are packed due to unbalanced intermolecular forces. Thus, there is an enhanced energy density at the surface of a liquid too. Both, surface energy and surface tension have the same dimension J/m^2.

In the case of a liquid droplet (or gas bubble) the surface energy (tension) is of isotropic character being constant over the entire surface and, thus, the surface energy and tension are identical. In contrast, at a crystal the surface energy shows anisotropically due to the differing atomic surface arrangement perpendicular to the crystallographic orientation. Its variance along the various crystal faces is expressed by γ_{hkl} with the subscripts *hkl* denoting the *Miller's indices*. Therefore, strictly speaking, for crystals the surface energy is not exactly equal to the surface tension. Defining the surface tension as the work necessary to increase the area by unit amount, it proves to be not more the same reaction on the work in all directions, especially in the cases of highly anisotropic crystal structures.

We have also to differ between the surface energy in one-component and multicomponent systems. As we introduced above, the adsorption by foreign atoms or molecules represents the contact of a given crystal with a multicomponent fluid phase. Then the free surface energy is a function on the presented foreign component mole fraction x_i and its value is extended by the sum of chemical potentials of the additives as

$$\gamma(x_i) = \gamma_0 + \sum_i \Gamma_i \mu_i \tag{4.4}$$

where γ_0 is the energy of a clean surface in a one-component system, μ_i the chemical potential of the *i*-component, and Γ_i the coupling factor representing the functional dependence of the surface energy on the concentration (mole fraction) of the *i*-component:

$$\Gamma_i = -\frac{x_i}{RT}\left(\frac{\partial \gamma}{\partial x_i}\right) \tag{4.5}$$

Substances (adsorbents) that reduce the surface energy ($\partial\gamma/\partial x_i < 0$) preferably accumulate there ($\Gamma_i > 0$) and are acting as *surface-active substances*. In the opposite case ($\partial\gamma/\partial x_i > 0; \Gamma_i < 0$) the substances behave *surface-passive* and are rather rejected (desorbed).

How to determine the surface energy? Its measurement in a liquid is simple due to its conformity with its surface tension, and a variety of techniques exist to measure liquid surface tension, e.g., by pendant drop method. However, determining the surface energy of a solid (crystal) is not nearly as simple because it cannot be directly measured. The surface energy values are calculated from a set of liquid-solid contact angles, developed by bringing various liquids in contact with the solid. For that, one must have prior knowledge of the surface tension values of the liquids that are used as well as to consider specific surface interactions, surface reactivities, and surface solubilities. Then *Young's equation* can be used

$$\gamma_S = \gamma_{SL} + \gamma_L \cos\theta \tag{4.6}$$

where γ_S is the overall surface energy of the solid, γ_{SL} the interfacial tension between the solid and the liquid, γ_L the overall surface tension of the wetting liquid, and θ the contact angle between the liquid and the solid. Due to the impracticality of obtainment of a universal set of liquids for use in testing solid surfaces, it is currently accepted that the most accurate values for surface energy of solids can be approximated from theoretical calculations.

Starting with the historical approach, *Turnbull* ascertained in 1950 during his study of maximum supercooling of metal droplets (see Section 5.1.1) that there is a correlation between the surface energy and enthalpy of fusion. Since Δh is a molar quantity it can be compared with the gram atomic surface energy between solid and liquid γ_m^{SL}, which may be defined as the free energy of an interface A containing Avogadro's number N_A. Assuming arbitrarily that the interface is one-atom thick and V_m is the molar volume then $A = N_A^{1/3} V_m^{2/3}$ and the molar interface energy becomes $\gamma_m^{SL} = \gamma_{SL} A = \gamma_{SL}\, N_A^{1/3} V_m^{2/3}$. It was observed that γ_m^{SL} is equal to about one-half of the latent heat of fusion per atom ($\gamma_m^{SL} \approx 1/2\ \Delta h$) for metals, and about one-third for water and organic compounds (*Turnbull's rule*). Correlating the factor of proportionality with the degree of nearest neighbors' occupation *Brice* proposed 1986 an analogous approximation as

$$\gamma_{SL} = \gamma_m^{SL}/A = (1 - w/u)\Delta h_m\, N_A^{-1/3} V_m^{-2/3} \tag{4.7}$$

where u is the number of nearest neighbors for an atom inside the crystal and w is the number of nearest neighbors for an atom on the surface ($w < u$).

Today, computational techniques, such as first *principles computations* based on density functional theory (DFT), provide the means to precisely control the surface structure and composition of a material. Principally, fundamental- and application-driven computational studies of surfaces in the literature are extensive. The standard structure used to calculate surface properties from first-principles is the *surface slab* (slide) – a supercell representing an infinite two-dimensional thin film oriented to expose the facet of interest, separated from periodic images by a large vacuum. The slab should be thick enough so that there is no interaction between opposite surfaces through the bulk, and the vacuum distance between slabs should be increased until there is no more interaction between adjacent slabs. For a converged, clean slab in vacuum, the surface energy can be defined as

$$\gamma_S = \frac{1}{2A}\left(E_{\text{slab}} - NE_{\text{bulk}}\right) \tag{4.8}$$

where A is the area of the surface unit cell, E_{slab} is the energy of the slab supercell, E_{bulk} is the bulk energy per atom, and N is the number of atoms in the surface slab. The 1/2 pre-factor accounts for the two surfaces of a slab. One of the overviews has been published by *Schultz* in 2021.

4.2 Gibbs–Thomson equation

Next we have to clarify on which influencing factor is the free surface energy during crystal growth processes. Sure, the answer lies in the size relation, more precisely, in the proportion between surface and volume share of a given crystal. Starting with the simplest example of a liquid droplet within a gas phase *Laplace* showed in 1805 that the surface (interface) energy of a curved interface between both liquid (L) and vapor (V) phases, γ_{LV} causes the pressure difference between both phases as

$$p_L - p_V = \frac{2\gamma_{LV}}{r} \tag{4.9}$$

where r is the droplet radius. Thus, the pressure in a small droplet ($r \to 0$) is always higher than that of the surrounding vapor. The ratio $2\gamma_{LV}/r$ becomes zero when the phase boundary is flat ($r \to \infty$) and $p_L = p_V$.

The *Gibbs–Thomson equation* uses this effect for the difference of the chemical potentials of the vapor μ_V and of a spherical droplet μ_L of a one-component system at constant temperature as

$$\mu_V - \mu_L = \Delta\mu = \frac{2\gamma_{LV}}{r}\Omega_L \tag{4.10}$$

where Ω_L is the molar volume in the condensed phase (note, in the name "Gibbs–Thomson" equation, "Thomson" refers to J. J. Thomson, not William Thomson/Lord Kelvin). The detailed derivation of the Gibbs–Thomson equation is given in the *Spec box 4.1.*

Spec box 4.1: Derivation of classic Gibbs–Thomson equation

At the presence of a liquid (L) droplet in thermodynamic equilibrium with a surrounding vapor phase (V) at constant pressure p and temperature T the differential of Gibbs equation (4.2) becomes

$$dG = \mu_V dn_V + \mu_L dn_L + \gamma_{LV} dA = 0 \tag{B4.1-1}$$

When the system is closed $n_L + n_V =$ const and $dn_L + dn_V = 0$ as well as $dn_L = -dn_V$ the eq. (B4.1-1) becomes

$$\mu_V - \mu_L = \gamma_{LV}\frac{dA}{dn_L} \tag{B4.1-2}$$

Setting for the interface area $A = 4\pi r^2$ and for the atom quantity within the droplet (the quasi volume part) $n_L = \frac{4\pi r^3}{3\,\Omega_L}$ where Ω_L is the molar volume, and differentiating $d(4\pi r^2)/d\left(\frac{4}{3}\pi r^3\right)$ the equation of Gibbs–Thomson becomes

$$\mu_V - \mu_L = \Delta\mu = \frac{2\gamma_{LV}\,\Omega_L}{r} \tag{B4.1-3}$$

Drawing the $\Delta\mu = f(r)$ function (Fig. 4.2a), one sees that the potential difference is exponentially decreasing with increasing radius. That means the influencing factor of the free surface energy increases with reducing (droplet) size due to the decisional factor of the curvature strength.

Figure 4.2a shows the functional dependences $\Delta\mu(r)$ according to eq. (4.10) of silicon, mercury, and water droplets within their own vapor phase. The smaller the droplet radius the higher is the potential difference. It can be concluded that the existence of a positive interfacial energy will increase the energy required to form small droplets with high curvature, and these droplets will exhibit an increased chemical potential. In Section 5.1, we will show that the $\Delta\mu$ is identical to the *supersaturation*.

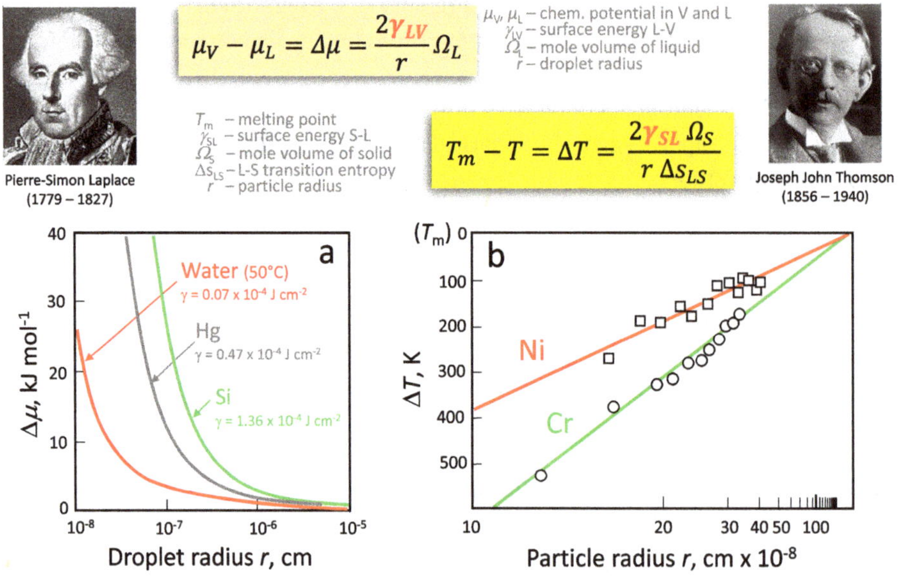

see e.g.: A. Gualda, in: Contrib. to Mineralogy and Petrology 174 (2019) 88

adapted from: R. Hashimoto et al., ISIJ Internat. 51 (2011) 1664

Fig. 4.2: Chemical potential and undercooling versus droplet and particle diameters influenced by surface energy of selected materials (*the portrait of P.S. Laplace and J.J. Thomson are public domains; with permission of Springer Nature (a); free to use from ISIJ Int.(b)*).

Let us assume a pure one-component solid (*S*) spherical phase of radius *r* situated in a large liquid phase (*L*) of the same component. Then at given temperature *T* eq. (4.10) is

$$\Delta\mu = \mu_L - \mu_S = (h_L - Ts_L) - (h_S - Ts_S) = \frac{2\gamma_{SL}}{r}\Omega_S \tag{4.11}$$

where γ_{SL} is the surface (interface) energy between the crystalline and melt phases and Ω_S the molecular volume in the solid phase. After conversion eq. (4.11) becomes

$$(h_L - h_S) = \Delta h_{LS} = T(\Delta s_{LS}) + \frac{2\gamma_{SL}}{r}\Omega_S \tag{4.12}$$

where Δh_{LS} is the transition enthalpy (latent heat of fusion) at transition (melting) temperature $T_{tr} = T_m$ (see Section 2.2) and Δs_{LS} the transition entropy. According to eq. (2.9) Δh_{LS} can be substituted by $T_m\ \Delta s_{LS}$. After some conversion of eq. (4.12) the depressed melting point of a microcrystal becomes

$$T = T_m - \frac{2\gamma_{SL}\Omega_S}{r\Delta s_{LS}} \quad \text{or rather} \quad T_m - T = \Delta T = \frac{2\gamma_{SL}\Omega_S}{r\Delta s_{LS}} \tag{4.13}$$

This variation of the classical Gibbs–Thomson equation describes the equilibrium melting point at which a solid “spherical” crystal of radius *r* has the same Gibbs energy as that of the surrounding large one-component liquid phase. In other words, a small crystal of equilibrium size has the same Gibbs free energy as that of the macroscopic liquid phase at a given *T*. Figure 4.2b shows the molecular dynamic (MD) simulated depression of the crystallization temperature *T* from the bulk melting point $T_m - T = \Delta T$ as function of the radius *r* of Ni and Cr particles. As can be seen the “undercooling” is enormous when the radius amounts only a few nanometers. Thus, in the sphere of *nanocrystal growth* the melting point is reduced to be considered in the respective phase diagram. This important relation we will meet again when the critical radius of a formatting nucleus is treated in Chapter 5.

It is clear that one cannot readily apply the classical Gibbs–Thomson equation (4.10) for isotropic cases to polyhedral crystal consisting of faces with their own specific surface free energy varying with the orientation of the surface relative to the crystal axes. Thus, an anisotropic crystalline solid will, in general, have a specific surface free energy, which varies with the crystallographic orientation. Here we will adapt some simple relations from the review of *Johnson* (1964) only. When denoting the surface orientation by the vectorial unit normal to the surface $\hat{n}$ then the specific surface free energy is $\gamma = \gamma(\hat{n})$. Considering a crystallite of a pure substance (*S*) in equilibrium with its surrounding fluid phase (*F*), and having therefore the equilibrium shape, compared to the droplet form of radius *r* the size of the crystallite is specified by the value $\mathfrak{R}\ (\hat{n})$ – a radial vector to the crystal surface for any direction $\hat{n}$. Therefore, eq. (4.10) is substituted by

$$\mu_F - \mu_S = \Delta\mu = \frac{2\gamma(\hat{n})}{\mathfrak{R}(\hat{n})}\Omega_S \tag{4.14}$$

where Ω_S is the molecular volume in the crystal. Equation (4.14) represents the *generalization of the Gibbs–Thomson equation* for crystals having full inversion symmetry in equilibrium with their surrounding fluid (mother) phase. According to this equation the chemical potential of the fluid in equilibrium with a small crystal is directly proportional to the specific surface free energy of any given surface orientation $\hat{n}$ and inversely proportional to the width of the crystal (the distance between the parallel planes tangent to the body at opposite sides) for the same orientation $\hat{n}$. The equation is valid for polyhedral equilibrium bodies as well as for smoothly curved equilibrium bodies. It is apparent that for a spherical equilibrium body with $\mathfrak{R}(\hat{n}) \to r$ and isotropic surface free energy ($\gamma(\hat{n}) \to \gamma_{FS}$) eq. (4.14) becomes analogous to eq. (4.10).

4.3 Equilibrium shape of crystals

In the following the anisotropic approach of the *equilibrium shape of crystals* will be somewhat more specified. To understand and simulate the equilibrium shape of a growing crystal proves to be one of the fascinating subjects of physical, especially crystallographic research, not least because of its proximity to art and aesthetics. Accordingly, it turns out to be a long-term subject of thinking, interpretation, and modeling until today. Even with the advent of the nanocrystal preparation it became significant due to the dominating role of the surface energy in relation to the volume part, respectively (see Fig. 4.2a and b). Therefore, there are several treatments, reviews, and textbooks. Here we will give a short insight into the basic features only.

In 1878, *Gibbs* proposed that a crystal will arrange itself such that its surface free energy is minimized by assuming a shape of minimum surface energy. According to eqs. (4.2) and (4.3), by exclusively considering the surface related term he defined

$$\Delta G_i = \sum_j \gamma_j A_j \quad \to \min \tag{4.15}$$

where γ_j is the surface (Gibbs free) energy per unit surface area A_j of the jth crystal face. ΔG_i represents the difference in energy between a real crystal composed of i molecules with a surface and a similar configuration of i molecules located inside an infinitely large crystal. This quantity is therefore the energy associated with the surface. From a crystallographic point of view the summation is performed along all *hkl* crystal plane (Miller) indexes, with the *hkl*-dependent surface energy and the specific surface area of each plane. Thus, the subscript j can be replaced by *hkl*.

In 1901, *Wulff* outlined his geometrical interpretation without proof whereupon in thermodynamic equilibrium the distance of each crystal face from an assumed point within the crystal is proportional to the corresponding specific surface energy of this face. According to this *Wulff's theorem*, depicted in Fig. 4.3a, one can draw vectors normal to all possible crystallographic faces from an arbitrary point P having the length h_j proportional to the corresponding surface energy γ_j. Then at the vector tip a perpendicular line is drawing representing the given crystal face f_j. After all vector-face sets are sketched, inside the crossing face lines representing the continuous sequence of minimum surface energies $\gamma_{\min}$ a closed polyhedral envelope (bolt blue contour) is formed to be equated with the equilibrium shape as it is sketched for the two-dimensional case in Fig. 4.3a. As can be seen, the perpendiculars f_3 of the longest vectors h_3 representing the largest surface energy γ_3 cannot contribute to the equilibrium contour by any face formation due to their too high γ_3 values. The geometrical interpretation confirms the relation

$$\frac{h_j}{\gamma_j} = \text{const} \qquad \text{or rather} \qquad \gamma_1 : \gamma_2 : \gamma_3 \ldots = h_1 : h_2 : h_3 \ldots \tag{4.16}$$

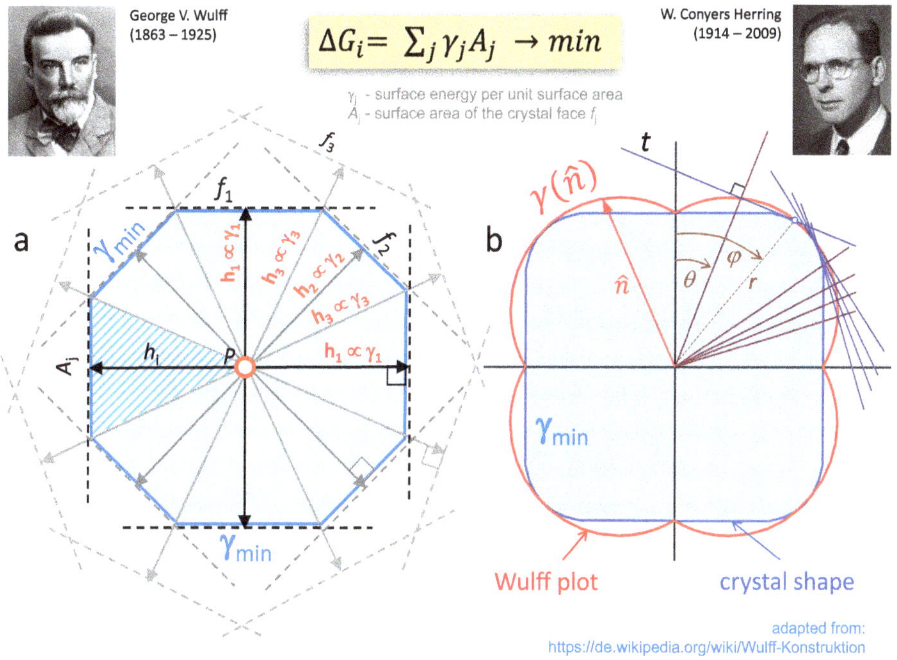

Fig. 4.3: Wulff's constructions to determine equilibrium crystal shape; a – starting from center and b – deducing from $\gamma(\hat{n})$ contour (*the portrait of G.V. Wulff is reproduced with permission of IUCr Newsletter 30 (2022) No. 2; the portrait of W.C. Herring is a public domain; the Wulff plot is in the public domain of Wikipedia*).

Corrected proofs of this theorem were subsequently given by *Hilton* (1903), *Liebmann* (1914), and, especially recommended, *von Laue* (1943). The simplest one of Hilton is demonstrated in the Spec box 4.2.

Spec box 4.2: The constancy of h_j/γ_j ratio

The volume of the crystal can be considered as a sum of the volumes of pyramids constructed on the crystal faces A_j with a common apex h_j in the arbitrary point P [hatched area in Fig. (4.3a)] as

$$V_{cr} = \frac{1}{3}\sum_j h_j A_j \tag{B4.2-1}$$

For a very small change in shape for a constant volume without varying the surface free energy, then

$$\delta V_{cr} = \frac{1}{3}\delta\left(\sum_j h_j A_j\right) = 0 \tag{B4.2-2}$$

which can be written, by applying the product rule, as

$$\sum_j h_j \delta A_j + \sum_j A_j \delta h_j = 0 \tag{B4.2-3}$$

whereas the second term must be zero if the volume remains constant. In other words, the changes in the heights of the various faces must be such that when multiplied by their surface areas the sum is zero. That means it remains

$$\sum_j h_j \delta A_j = 0 \tag{B4.2-4}$$

The condition of surface energy minimization [see eq. (4.15)] is for a given volume when

$$\delta\left(\sum_j \gamma_j A_j\right)_{V_{cr}} = \sum_j \gamma_j \delta(A_j)_{V_{cr}} = 0 \tag{B4.2-5}$$

because surface free energy as an intensive property does not vary with volume. Considering eq. (B4.2-5) and employing a constant of proportionality λ, eq. (B4.2-4) becomes

$$\sum_j \left(h_j - \lambda\gamma_j\right)\delta A_j = 0 \tag{B4.2-6}$$

The change in shape $\delta(A_j)_{V_{cr}}$ must be allowed to be arbitrary, which requires that $h_j = \lambda\gamma_j$ or $h_j/\gamma_j = \lambda$, a constant being equal for all face areas A_j and relating to the crystal volume only.

Although such a simple design of the so-called *Wulff's plot* shown in Fig. 4.3a is of sufficient comprehensibility, even for beginners, it does not turn out to be quite correct. Such plot corresponds to the zero temperature only. At finite temperature its sharp corners tend be rounded. In 1953 *Herring* introduced a more correct chronology of Wulff's construction sketched in Fig. 4.3b. To begin, a polar plot of surface energy as a function of orientation is made (red contour in Fig. 4.3b). This is known as the "gamma plot" and is usually denoted as $\gamma(\hat{n})$ contour (or Wulff plot) with $\hat{n}$ the surface normal belonging to a particular face. The second part is the Wulff construction itself in which the gamma plot is used to determine graphically, which crystal faces will be present. Toward this end, lines from the origin to every point on the gamma plot are drawn. Then, perpendiculars ***t*** to the normals are drawn toward each point, where it intersects, as tangent, the gamma plot by the radius r from the center. This point is now determined by two angles, i.e., the tangent related θ between y-axis and normal $\hat{n}$ and φ between y-axis and r. The inner envelope of all these ***t***-lines, representing de facto the crystal faces, forms the equilibrium shape of the crystal often named $\gamma(\theta,\varphi)$ plot (blue contour in Fig. 4.3b). In this way quasi-rounded crystal regions are recognized tangentially. The first step, however, i.e., the construction of $\gamma(\hat{n})$ contour proves to be somewhat extensive requiring the integration over the whole *hkl*-belonging face energies of the given crystal structure (see related review papers and textbooks).

Another decisional insufficiency of Wulff's energetic minimum shape is that it is exclusively based on *atomically smooth*, so-called *singular faces*. However, even faces with low surface energy tilted from a singular face by a small angle are not flat on the atomic scale but are stepped. Such steps and terraces that exhibit areas are named *vicinal faces*. Figure 4.4 shows a macroscopically quasi-rounded crystal contour (dashed line) with normal vectors $\hat{n}$, which on microscopical scale is partitioned into a central singular and sideward vicinal faces sloped by the small angle Θ. The specific surface energy of such a stepped face is the sum of the surface energy of the (atomically smooth) terraces γ_{j0} (being identical to those of the singular central face) and that of the steps ω_{jst}/a, with ω_{jst} the work required to create a unit length of the step and a the step height. Assuming a constant terrace width and a small slope angle the specific surface energy of such vicinal face is

$$\gamma_{\mathrm{vic}}(\Theta) = \frac{\omega_{jst}}{a}\sin\Theta + \gamma_{j0}\cos\Theta \tag{4.17}$$

and for the vicinal face being symmetric to the first one but tilted by an opposite angle $-\Theta$

$$\gamma_{\mathrm{vic}}(-\Theta) = -\frac{\omega_{jst}}{a}\sin\Theta + \gamma_{j0}\cos\Theta \tag{4.18}$$

The polar diagrams of the surface energy for a simple-cubic crystal (at $T = 0$) are sketched by *Chernov* (1984) in the Fig. 4.5. The $\gamma_{\mathrm{vic}}(\pm\Theta)$ functions from eqs. (4.17) to (4.18) in orthogonal coordinates are represented in Fig. 4.5a. As can be seen the $\gamma_{\mathrm{vic}}(\pm\Theta)$ function shows discontinuities of the first derivatives ($d\gamma_{\mathrm{vic}}/d\Theta$) for $\Theta \leq 0 \leq \Theta$ at $\Theta = 0, \pm\pi/2, \pm\pi$ named *singular points*. In Fig. 4.5b the same function is plotted in 2D polar coordinates consisting of circular segments and singularities at $\Theta = 0$, $\pi/2$, and $3/2\pi$. Figure 4.5c represents the three-dimensional depiction taking into account first neighbor interactions only. The graph consists of 8 spherical segments with 6 singularities coinciding with 6 atomically flat (singular) faces. Finally, Fig. 4.5d shows the polar diagram of specific surface energy considering the first-neighbor and second-neighbor interactions together. In a simple cubic crystal the bonds of the second neighbors are directed at an angle of $\pi/4$

Zinc micro crystal with singular basal plane and vicinal faces around it grown from vapor

Silicon LPE film grown on a convex curved Si substrate with central (111) singular facet and vicinal faces around it

Ch. Nanev and D. Ivanov, J. Crystal Growth 3 (1968) 530

E. Bauser and H. Strunk, J. Crstal Growth 69 (1984) 561

$$\gamma_{vic}(-\Theta) = -\frac{\omega_{jst}}{a}\sin\Theta + \gamma_{jo}\cos\Theta$$

$$\gamma_{vic}(\Theta) = \frac{\omega_{jst}}{a}\sin\Theta + \gamma_{jo}\cos\Theta$$

adapted from: A.A. Chernov, Modern Crystallography III. Crystal Growth (Springer Vlg. 1084)

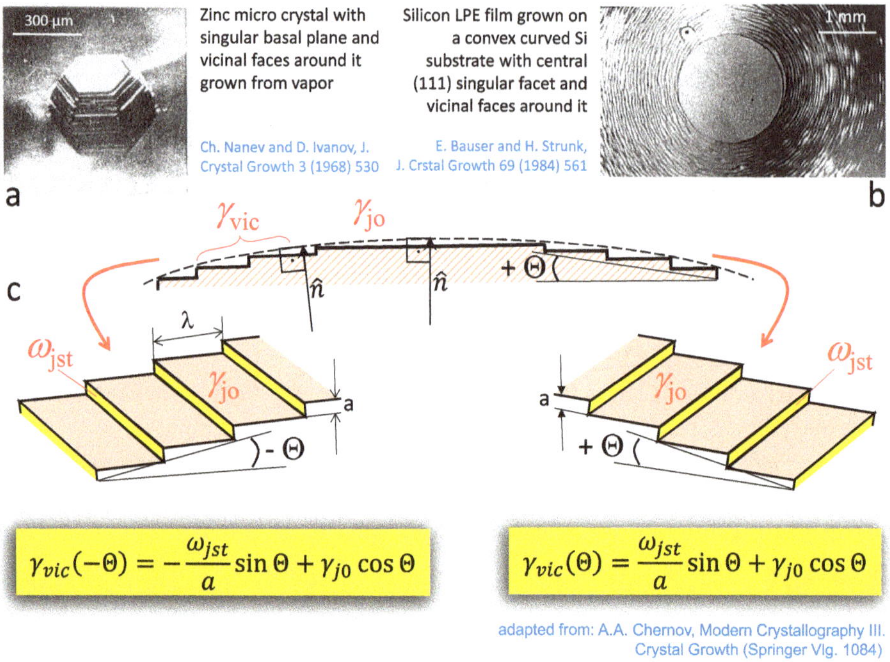

Fig. 4.4: Images of vicinal (stepped) faces at a Zn micro crystal (a) and silicon LPE film (b). Determination of the surface energy of vicinal faces γ_{vic} deviated from singular face by a small angle θ (c) (*with permission of Elsevier (a and b) and Springer Nature (c)*).

with respect to the first-neighbor bonds. Therefore, the polar diagram $\gamma(\Theta)_2$ belonging to the much weaker second-neighbor bonds is inscribed in the first one $\gamma(\Theta)_1$ and rotated with respect to it by an angle of $\pi/4$. The sum of both functions along the outmost contour $\gamma(\Theta)_1 + \gamma(\Theta)_2$ generates additionally to the (100) singularities shallower singularities of (110) faces. Inside of all singular minima comes up the polygon of equilibrium surface energy (sketched in blue). In the three-dimensional case a polyhedron consisting of the equivalent plane families {100} and {110} arises. Typical minimum values of $\gamma(\Theta)$ at crystal–melt and crystal–vapor interfaces are in the ranges 0.1–1.0 J/m^2 and 1.0–10 J/m^2, respectively. In solutions the $\gamma(\Theta)$ function is somewhat modified by adsorption effects of the solvent atoms or molecules at the growing interface and decreases or increases if the solvent is surface-active or -passive, respectively (see Section 4.1). Of course, taking into account the bond states and more distant neighbors in the 7 crystal systems, including 14 Bravais lattices, more differentiated and complicated polar diagrams with related surface equilibrium shapes will result. Some related suggestions even in the case of nanocrystals are given in the excellent book chapter of *G. Guisbiers* and *M. Jose'-Yacaman* (2018).

Thus, compared to Wulff's plot now the depiction takes the following correct chronology: one begins by creating an accurate polar plot of the surface free energy

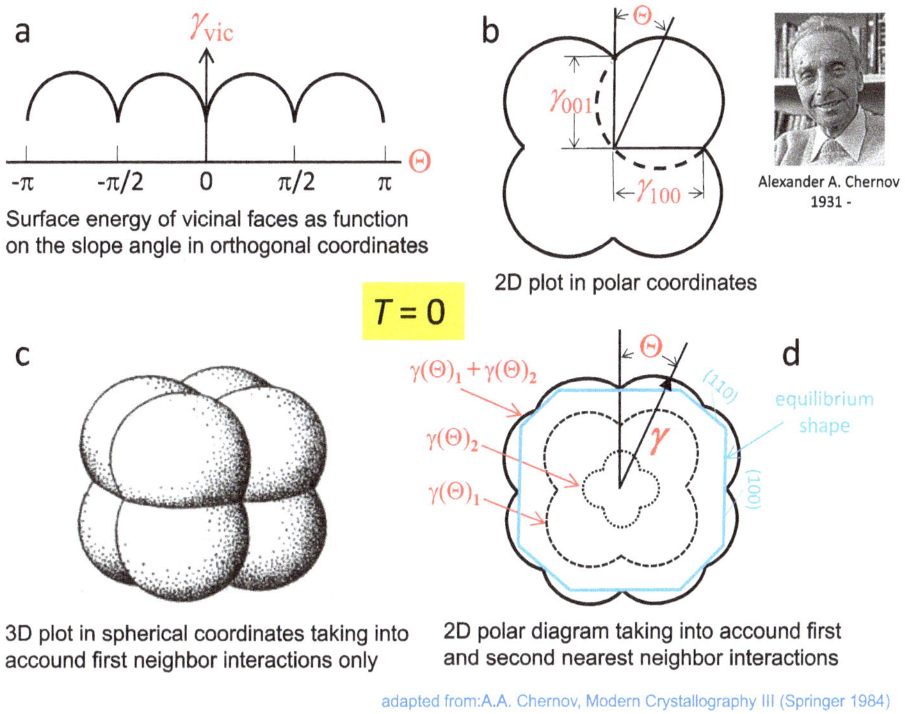

Fig. 4.5: Orthogonal (a) and polar diagrams (b and c) of surface energy of a simple cubic crystal considering first (b, c) and second nearest neighbor interactions (d) (*the portrait of A.A. Chernov is reproduced from J. of Lawrence Livermore National Laboratory with its permission under the license CC BY-NC-SA 4.0; with permission of Springer Nature (a–d)*).

as a function of orientation angle, i.e., the $\gamma(\Theta)$ function often called *surface energy rosette*, and draws a perpendicular inside plane contour through the tip of each minimum. The equilibrium shape is then designed by the interior envelope of these planes (e.g., in 2D lines).

It is important to point out that in reality the surface free energy-determined equilibrium crystal shape is mostly only realized at low growth temperatures (from vapor or solution) and very small crystals, so-called crystallites or *nanocrystals*, during and immediately after their nucleation. First let us have a view on the effect of temperature. Representative Wulff plots and crystal shapes for the simple-cubic lattice-gas model are shown in Fig. 4.6. Starting at $T = 0$ the equilibrium shape consists still of flat faces being more or less smooth and so are the steps on them. With increasing temperature, however, the thermal fluctuations become important and the steps become more and more rough. Heavy dots on the 2D crystal shapes in Fig. 4.6 mark the singularities where the flat faces meet the smoothly curved regions. At first the corner edges (111) (having the highest free surface energy of simple cubic structure) are rounded and their flatness disappears entirely at T_{R1}. Thereafter the (110) edges are vanished at T_{R2}. These moments are characterized by tending the step Gibbs free energy to zero. Finally, the crystal shape is everywhere smoothly curved and spherical symmetry is adapted as $T \approx T_c$ the *critical temperature* meaning that the whole crystal surface consists entirely of rough character, even as T approaches melting temperature T_m. However, T_c must not yet coincide with the melting temperature but may be quite lower (more details will be given in the lecture part II on growth kinetics). Some corresponding images of various near-equilibrium microcrystals with dimensions not more than 100 µm are added in Fig. 4.6. All of them were grown from vapor or solution at temperatures $0 < T < T_R$, T_{R1}, T_{R2}, and T_c, respectively. On the right side, a modeled simple-cubic crystal at corner viewed from the {111} direction is added (in the used *Ising model* the temperature is quasi-quantified in the form of fluctuation size). It shows clearly the T-oscillation-driven rounding effect of the (111) and (110) edges due to the introduced step disorganization.

With increasing crystal size the free surface energy becomes more and more ineffective and many further environment factors, such as adsorptions, solvents, surfactants, impurities, surface reconstructions, dislocations, stacking faults and twins, devi-

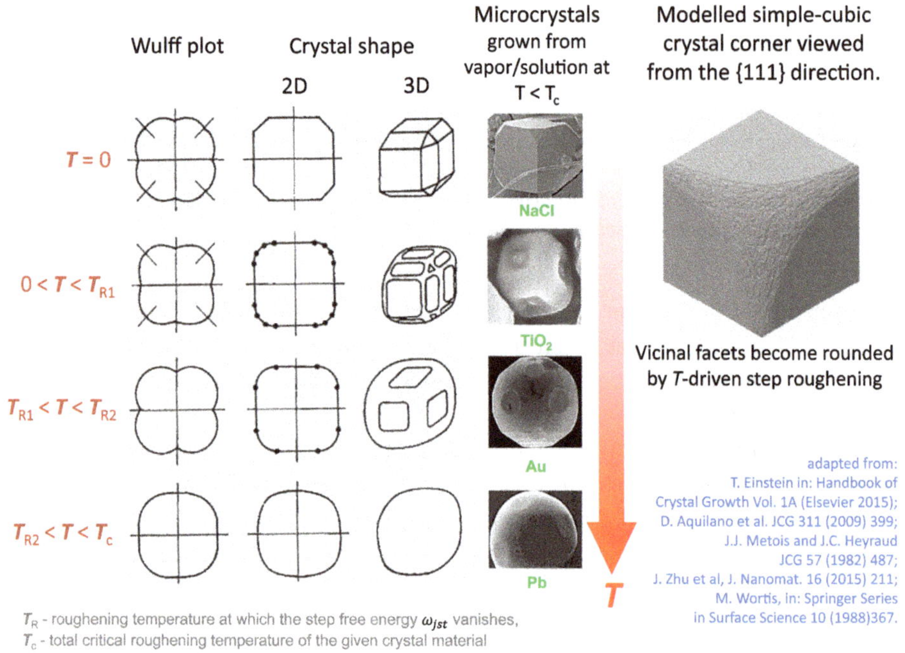

Fig. 4.6: Edge rounding due to effect of temperature roughening (*with permission of Elsevier, Springer Nature, and via open accesses by MDPI (TiO_2)*).

ation from stoichiometry, degree of supersaturation or supercooling, crystallization velocity, morphological instabilities, and even the structural state of the fluid phase (degree of association) affect its value. Best practical studies of the equilibrium crystal shapes are possible at high-purity vapor growth by sublimation-condensation in closed ampoules or solution growth under ultraclean conditions. Also in the case of epitaxial processes at the stage of 3D island growth the impact of process parameters on the size ratio of the equilibrium faces proves to be well observable. Figure 4.7 shows habit variations of selected crystallites during early growth stage from vapor and solution as a function of various parameters that affect it. Figure 4.7a presents an image of a needle-shaped KDP (KH_2PO_4) mesocrystal grown from aqueous solution the {100} side faces of which are retarded, obviously due to an affecting *surfactant*. Such effect takes place when ethanol is added into the water, for example. Whereas the four pyramidal {110} faces are terminated with a layer of K^+ ions only, the {100} faces are terminated with both K^+ and $H_2PO_4^-$ ions. Then, the growth suppression of {100} faces for added ethanol is due to the formation of strong saturating hydrogen bonds between the hydroxyl groups of the alcohols and the $H_2PO_4^-$ ions reducing the free surface energy.

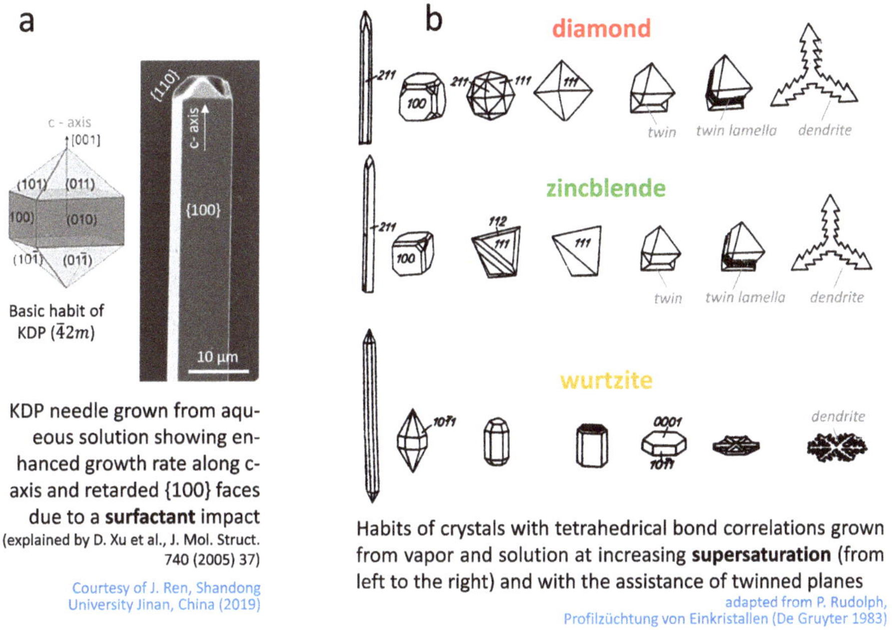
a
c - axis
[001]
(101)
(011)
(100)
(010)
(101̄)
(011̄)
Basic habit of KDP (4̄2m)
{110}
c- axis
{100}
10 µm
KDP needle grown from aqueous solution showing enhanced growth rate along c-axis and retarded {100} faces due to a **surfactant** impact
(explained by D. Xu et al., J. Mol. Struct. 740 (2005) 37)
Courtesy of J. Ren, Shandong University Jinan, China (2019)
b
diamond
211
100
111
twin
twin lamella
dendrite
zincblende
112
wurtzite
0001
dendrite
Habits of crystals with tetrahedrical bond correlations grown from vapor and solution at increasing **supersaturation** (from left to the right) and with the assistance of twinned planes
adapted from P. Rudolph, Profilzüchtung von Einkristallen (De Gruyter 1983)

Fig. 4.7: Habit variations at growth from vapor and solution depending on various affecting parameters (*courtesy (a) and with permission of De Gruyter (b)*).

Already in 1976 *Stroitjeljev* pointed to the variety of polyhedral habits of typical semiconductor crystal structures as a function of the degree of association, supersaturation, and impureness of the starting fluid phase (vapor and solution) as well as on the imperfectness, such as dislocations and twins, of the growing crystal. The mesocrystal habits then observed and sketched still remain useful today (see Fig. 4.7b). With increasing supersaturation (from left to right) both the twinning probability and dendrite formation are increasing. As a result the formation of plate-like morphologies is typical. Nowadays, these early studies help understand and control the habits of nanocrystals. All nanomaterials share a common feature of large surface-to-volume ratio, making their surfaces the dominant player in many physical and chemical processes. This also applies to the nucleation and growth process of epitaxial layers on heterogeneous or structured substrates. For instance, at the growth of GaN on patterned sapphire substrates the facet evolution of the growing crystallites before their coalescence is more and more studied by consideration of *kinetic Wulff plots* under different temperatures, vapor composition ratios, undoped and doped conditions. For that the $\gamma(\theta,\varphi)$ function of the usual Wulff plot (see above) is substituted by a semi-empirical kinetic function $v(\theta,\varphi)$ with v the analyzed growth velocity of the appeared planes. As a result, a methodology can be developed to apply the experimentally determined v-plots to the interpretation and design of evolution dynamics in nucleation and island coalescence.

4.4 Selected effects of surface energy on crystal growth processes

4.4.1 Growth angle at crystal pulling from the melt

The surface energy plays a decisional role at bulk crystal growth methods, especially when free pulling from the melt is used, like at Czochralski (CZ), TSSG, floating zone (FZ), edge-defined film-fad growth (EFG), Stepanov, and micro-pulling down (μ-PD). Due to the absence of contact between the growing crystal and any crucible the solidification occurs on a *liquid meniscus* the shape of which is controlled by surface tension-driven capillary forces. As is observed, there is a strong relationship between the shape of the meniscus cross section and the growing crystal form as well as the meniscus stability and crystal diameter constancy.

Considering the capillary conditions at the solid–liquid interface during pulling processes from the melt an important dynamic parameter to be controlled is the meniscus height h_m along the crystal pulling axis z. This value is obtained by integrating the *Laplace equation*

$$\gamma_{LV}\left(\frac{1}{r_1}+\frac{1}{r_2}\right)=z\rho_l g \tag{4.19}$$

where γ_{LV} is the interface tension between the liquid (meniscus) and vapor (surrounding gas), r_1 and r_2 the principal two radii of curvature of the meniscus, z the vertical coordinate, ρ_l the melt density, and g the gravitational acceleration. Unfortunately, the mathematical description of the meniscus shape by eq. (4.19) is a second-order differential equation, which cannot be integrated analytically. Several analytical approximations have been reported in the literature. One of the widely used meniscus relations for the Czochralski case was given by *Boucher and Jones* in 1980 as

$$h_m = a\left[\frac{1-\sin\varphi_0}{1+(a/r\sqrt{2})}\right]^{1/2} \tag{4.20}$$

where $a=\sqrt{2\gamma_{SL}/\rho_l g}$ is the *capillary constant* with γ_{sl} the solid–liquid interfacial tension, r the crystal radius, and φ_0 the so-called *growth angle* between the meniscus slope and the vertical (i.e., the crystal mantle) at $z = h_m$. To obtain a uniform crystal cross section with r = const, a stable given meniscus height h_m must be obtained (a small correction of the denominator was given by *Johansen* in 1994). Looking at the Czochralski sketch on the left side above in Fig. 4.8 at first sight one might suppose that φ_0 should be zero as a requirement for pulling of crystals with constant diameter because then the meniscus turns into the vertical at the solid–liquid interface. However, there is a crystal–liquid–vapor triple contour the equilibrium of which at the melting point is determined by the interaction of the three interface energies γ_{SL}, γ_{SV}, and γ_{LV} as magnified in the center of Fig. 4.8. This equilibrium forces the growth angle φ_0 between crystal mantle and meniscus tangent to be almost somewhat greater than zero. Thus, φ_0 is a characteristic thermodynamic parameter of each material in contact with its melt under a given surrounding atmosphere (vacuum or inert gas).

The exact determination of φ_0 at the distinguished planes {hkl} of each crystal is complicated and, thus, often not correctly replaced by the wetting angle θ_{SL} because of the assumption of complete wetting of a crystal face with its own pure melt, so that the growth angle is erroneously taken as $\varphi_0 = 0$. However, this would mean that the diameter of most crystal substances increases continuously during the whole pulling process. Actually, there are only few materials showing complete melt wetting and, thus, a zero growth angle, like Cu, $LiNbO_3$, $LiTaO_3$, and ice, for example. However, most materials do not wet their own melt completely and exhibit a growth angle $\varphi_0 > 0$. Selected values are compiled in the table in Fig. 4.8. A phenomenological approach to this value by considering the interactions between the three participating interface energies at the crystal-melt-vapor contour, assuming that at any point of the triple line (meniscus-interface periphery) $\Sigma\gamma_{ij} = 0$, was given by *Voronkov* (1963) as

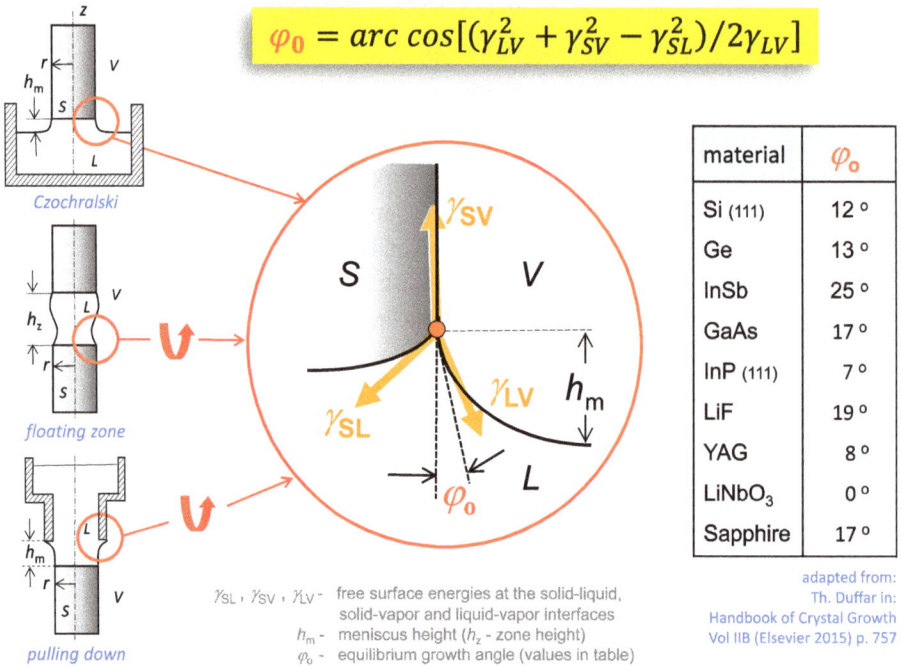

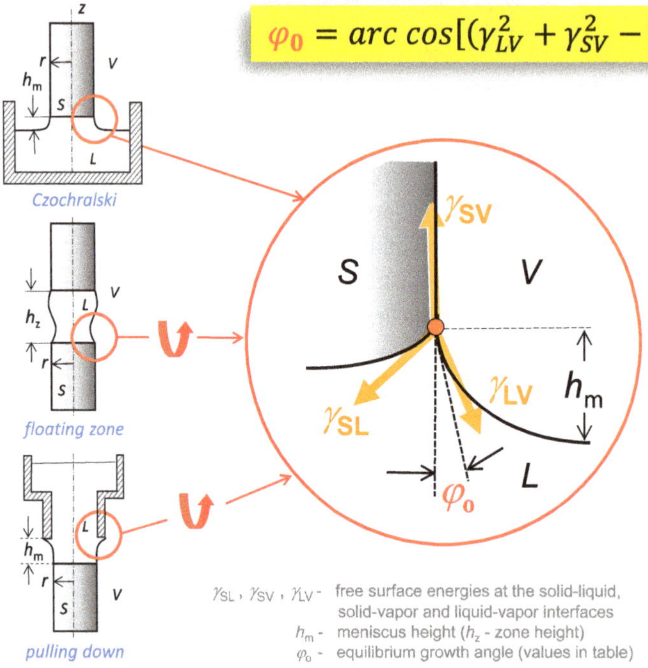

material	φ_0
Si (111)	12 °
Ge	13 °
InSb	25 °
GaAs	17 °
InP (111)	7 °
LiF	19 °
YAG	8 °
$LiNbO_3$	0 °
Sapphire	17 °

adapted from:
Th. Duffar in:
Handbook of Crystal Growth
Vol IIB (Elsevier 2015) p. 757

Fig. 4.8: The equilibrium growth angle φ_0 occurs at the crystal–liquid–vapor triple contour of melt growth methods with meniscus formation determined by the interaction of the three interface energies γ_{SL}, γ_{SV}, and γ_{LV} (*with permission of Elsevier (table)*).

$$\varphi_0 = \text{arc cos}\left[\left(\gamma_{LV}^2 + \gamma_{SV}^2 - \gamma_{SL}^2\right)/2\gamma_{LV}\right] \tag{4.21}$$

Table 4.1 shows γ_{SL}, γ_{SV}, and γ_{LV} values of selected crystal materials by taking into account the crystallographic orientation (Miller indices).

To come back for a moment to the effect of the capillary constant a, it determines the maximum height of a free liquid zone $h_z^{\max}$ between two solid bodies as $\sqrt{\gamma_{SL}/\rho_l g}$. This is of significance for crucible-free floating zone (FZ) processes whereupon the allowed zone height is decisional determined by the given γ_{sl} value. For instance, whereas at silicon $h_z^{\max}$ reaches 1.7 cm, in the case of GaAs it yields 8 mm only.

4.4.2 Faceting and ridge formation at melt growth

The mainly used melt-growth techniques are producing crystals of cylindrical shape. Thus, also the adjacent melt at the solid–liquid interface shows a round cross section. In fact, this should mean that also the peripheral contact line between the three phases, solid, liquid, and vapor, or rather the solid container wall (*triple phase line*) is circular. However, this proves to be only true from the macroscopic perspective. Mes-

Tab. 4.1: Interface energies γ_{SL}, γ_{SV}, and γ_{LV} of selected crystalline materials in (J/m^2) by considering the crystal face orientation (*with permission of Wiley, Elsevier, and Springer Nature*).

Material	γ_{SL} **(hkl)**	γ_{SV} **(hkl)**	γ_{LV}
Cu	1.70 (100)	1.35	0.14
Ag	1.20 (111)	1.17 (111)	0.90
Si	1.08 (111)	0.72 (111)	0.57
Ge	0.76 (111)	0.63 (111)	0.62
GaAs	0.50 (111)	1.05 (111)	0.45
		1.06 (100)	
CdTe	0.62 (111)	0.58 (111)	0.23
		0.85 (100)	
$LiNbO_3$	0.12 (0112)	0.09	0.18

Note that the values are averaged from theoretical and experimental literature data. Fizitcheskaja Khimija Poverkhnostnykh Javlenij v Rasplavakh (Nauka Dumka, Kiev 1971); S. Adachi, Properties of Group -IV, III-V and II-VI Semiconductors (J. Wiley, 2005); Th. Duffar in: Handbook of Crystal Growth Vol. IIB (Elsevier 2015) p. 757; G. Guisbiers, G. Abudukelimu, J. Nanopart Res. 15 (2013) 1431; R. Shetty et al., J. Crystal Growth 100 (1990) 51 and 58; P. Reiche et al., J. Crystal Growth 108 (1991) 759.

oscopically, in the vicinity of the triple-phase line the lateral crystalline surface can vary from a circular one due to the possible presence of *facets* formed by most closely packed atomic smooth (singular) crystal faces (see Section 4.3). This is the case when the normal of a singular face forms a favorable angle to the pulling axis and the related facet slope coincides with the peripheral curvature of the melt–solid interface, as sketched in Fig. 4.9 (left above). For example, at growth of crystals with diamond and zincblende structure in <001> direction there are four {111} faces crossing the four radial <110> directions showing an angle of 54.7° with the growth axis. As a result, at the triple-phase line four {111} facets are formed inclined towards the horizontal (melt surface) by an angle of 35.3°. This occurs because singular faces must draw back from the melting point isotherm into the colder region in order to become a higher undercooling for generation of the necessary rate of two-dimensional nucleation (around 3–5 K for dislocation-free {111} facets at high purity silicon). In comparison, on nonfacetted (atomically rough) surfaces, atoms can be added singly without the need for nucleation (let us assume that in Fig. 4.9 the main crystalline surface is of such configuration). The situation is similar for numerous other materials even when the crystal structure differs from cubic system. For instance, $LiNbO_3$ of trigonal crystal system shows the main singular faces (00.1) and {10.2}. As a result, at Czochralski growth along the usually used [12.0] direction three {10.2} facets appear on the triple-phase line angled in 65.12° (2×) and 32.75° (1×) to the melt surface. Thus, before starting a crystal pulling experiment it is recommended to sketch the surface equilibrium polyeder of the given substance in relation to the pulling axis in order to determine the slope of the singular faces toward the melt surface and their

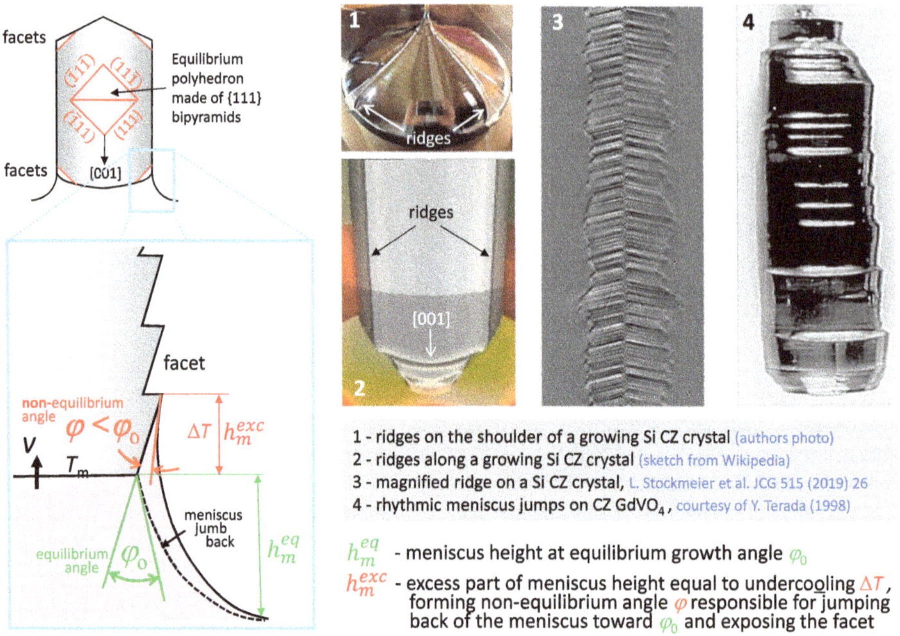
facets
Equilibrium polyhedron made of {111} bipyramids
facets
[001]
facet
non-equilibrium angle
$\varphi < \varphi_0$
ΔT
h_m^{exc}
V
T_m
meniscus jumb back
equilibrium angle
φ_0
h_m^{eq}
1
ridges
2
ridges
[001]
3
4
1 - ridges on the shoulder of a growing Si CZ crystal (authors photo)
2 - ridges along a growing Si CZ crystal (sketch from Wikipedia)
3 - magnified ridge on a Si CZ crystal, L. Stockmeier et al. JCG 515 (2019) 26
4 - rhythmic meniscus jumps on CZ $GdVO_4$, courtesy of Y. Terada (1998)
h_m^{eq} - meniscus height at equilibrium growth angle φ_0
h_m^{exc} - excess part of meniscus height equal to undercooling ΔT, forming non-equilibrium angle φ responsible for jumping back of the meniscus toward φ_0 and exposing the facet

Fig. 4.9: Ridges formation at atomically smooth facets caused by the interplay between nonequilibrium meniscus angle φ and equilibrium growth angle φ_0 illustrated by surface images of selected crystals as-grown from the melt (*public domain of Wikipedia (2) and with permission of Elsevier (3)*).

possible appearance as facets on the triple line (as it is sketched in the figure left above).

Figure 4.9 left shows the magnified sketch of meniscus dynamics along a facet. The melt–solid interface is assumed as an even plane growing with constant velocity v. A facet with a certain angle of inclination is added at the periphery. As a first approximation its lateral extension is proportional to its required growth undercooling ΔT differing from that of the rest of interface. At the early stage of facet generation the meniscus is stable attached to it by the constant equilibrium growth angle φ_0. However, with increasing facet length the meniscus onto it is more and more raised by an exceeding meniscus height h_m^{exc} and, thus, the contact angle φ is reduced in comparison with φ_0. Now, it is often the case, that even before the nucleation on the facet starts, the meniscus jumps back from the nonequilibrium situation $\varphi < \varphi_0$ to the stable one $\varphi = \varphi_0$. As a result, the facet area becomes exposed. In the course of the pulling process the rhythmic rising and sliding down of the meniscus forces a trace of *ridge-like protrusions* arranged in vertical seams along the crystal surface as shown by some exemplary crystal images 1–4 in Fig. 4.9.

From a practical point of view the genesis of ridges is of ambivalent appearance. One unfavorable feature proves to be the probability of spontaneous nucleation when the quasi-undercooled facet is overflowed by the melt during crystal rotation. This can happen at Czochralski growth with rotation, particularly, in the stage of shoulder growth. Furthermore, the back reflection of heat at the mirroring facet can provoke temperature inhomogeneities along the meniscus periphery, often resulting in spirally growing crystal shapes. Many studies have been carried out to find out the best experimental conditions for minimization of the facet area and related ridges sizes. One measure is the increase of the local temperature gradient by proper modeling-based heat zone engineering. Recently some patents were published claiming an azimuthally alternating insulation between growth container and heater well-adjusted to the facet positions at VB and VGF growth of semiconductor compounds. On the other hand, the presence of ridges is a visual evidence for crystal perfectness as far as this is in situ observable. For instance, at CZ or FZ methods both the correct number and arrangement of the facets along the propagating melt–solid interface indicate the single crystalline growth.

Faceting becomes increasingly problematic the stronger the binding force and anisotropy of the given crystal structure. Particularly, this mostly occurs with the CZ growth of substances with ionic bonds, such as oxides. Often the morphology of such crystals is not able to follow the cylindrical shape but split the mantel into numerous adjacent facets. Such crystal habit is named "idiomorphic". Figure 4.10 presents the images of some oxide crystals of idiomorphic forms. Typically for the strengths of ionic bonds and distinct anisotropic crystal lattice there is the distinct tendency to expose their most closely packed planes with lowest free surface energy, like in the case of melilite-type borate $Bi_2ZnB_2O_7$, bismuth silicon oxide $Bi_{12}SiO_{20}$, and langasite $La_3Ga_5SiO_{14}$ (a–c) with high nonlinear optical, photorefractive, or piezoelectric quality, respectively. The image of a Czochralski-grown Bi_2Te_3 crystal is added (d). This semiconductor is characterized by its high thermoelectric efficiency. However the pulling of cylindrical crystals from the melt proves to be almost impossible due to the trigonal lattice system with pronounced layered character. Perpendicular to the *c*-axis a quintuple layering with only very weak binding forces between the layers takes place. Hence, at the pulling and even simultaneous rotation of such crystals perpendicularly to the *c*-axis always oval cross sections (Fig. 4.10) or ribbon-like crystal shapes are obtained. In the last case, the bordering surfaces are two opposing flat mirror-like *c*-facets being equivalent to the easy cleavable basal plane with lowest surface free energy. On the other hand, attempts to grow along the *c*-axis always fail by a self-reorientation into directions markedly tilting from *c*. Therefore, it became clear that such crystals can be better grown by horizontal or vertical Bridgman techniques.

Finally, the crystallization from solutions or vapor phase is almost free of any shape constraint. As a result the crystal habitus most likely approaches the equilibrium one. Usually, the polyhedral shapes are limited by planes of low free surface energies as it is presented by selected solution-grown crystals in Fig. 4.10e-g (note, the

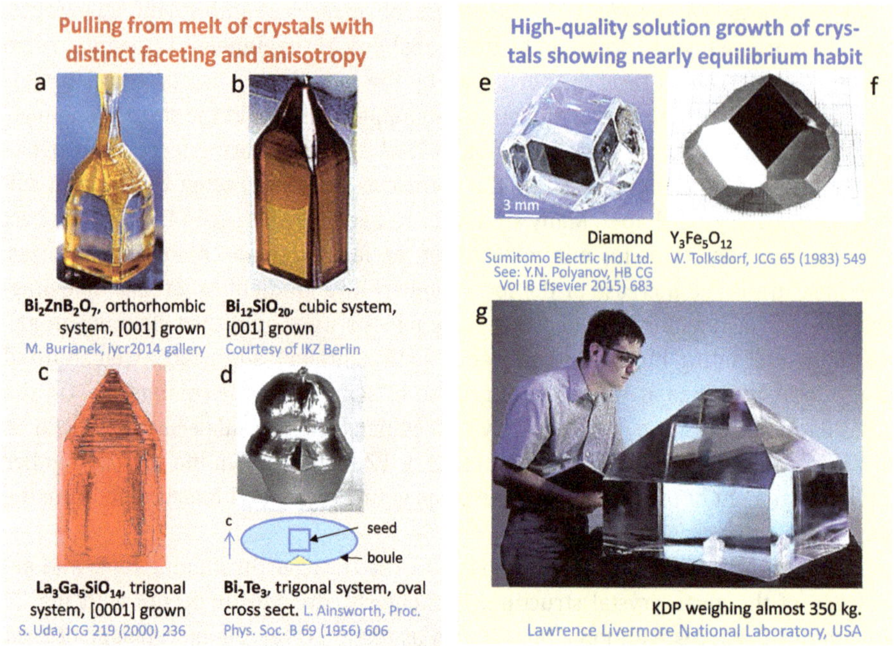

Pulling from melt of crystals with distinct faceting and anisotropy
a
b
$Bi_2ZnB_2O_7$, orthorhombic system, [001] grown
M. Burianek, iycr2014 gallery
$Bi_{12}SiO_{20}$, cubic system, [001] grown
Courtesy of IKZ Berlin
c
d
seed
boule
$La_3Ga_5SiO_{14}$, trigonal system, [0001] grown
S. Uda, JCG 219 (2000) 236
Bi_2Te_3, trigonal system, oval cross sect. L. Ainsworth, Proc. Phys. Soc. B 69 (1956) 606
High-quality solution growth of crystals showing nearly equilibrium habit
e
f
3 mm
Diamond
Sumitomo Electric Ind. Ltd. See: Y.N. Polyanov, HB CG Vol IB Elsevier 2015) 683
$Y_3Fe_5O_{12}$
W. Tolksdorf, JCG 65 (1983) 549
g
KDP weighing almost 350 kg.
Lawrence Livermore National Laboratory, USA

Fig. 4.10: Facet appearance on various crystals grown from melt and solution (*Creative Commons Attribution License (a); with permission of IKZ Berlin (b), Elsevier (c,e,f), IOP (d) and Lawrence Livermore National Laboratory (g)*).

shown diamond crystal was grown by using high-purity Fe–Co alloy as the solvent with added Ti and Cu to get nitrogen and minimize the formation of TiC, respectively). Of course, as it was already mentioned in Section 4.3, with increasing crystal size the free surface energy becomes increasingly ineffective and many environment factors are affecting its value. In solutions the degree of purity and the presence of additives (mineralizers) can influence the crystal shape, as was already demonstrated by the formation of needle-like KDP crystals when ethanol was added to the aqueous solution demonstrated in Fig. 4.8.

4.4.3 Surface energetic effects at epitaxy

In advanced epitaxial processes, the state of the surface plays an eminent role. Already before the deposition is started the structural feature, purity, and preparation quality of the substrate are decisional factors. First its *reconstruction*, responsible for minimization of the free surface energy, may influence the nucleation and initial layer growth modes onto it. Then, very thin films thereon possess the de facto substrate surface features as a whole. On heterogeneous (misfitting) substrates *strained layer situations* and *atomic ordering effects* may take place within the first few monolayers of multicomponent depositing materials. As a result the physical properties of the thin films are modified relative to those of the bulk state. Finally, the universal force of minimization of the interfacial energy can lead to *self-assembling* (self-organizing) arrangements of the nucleated islands into periodic nanopatterns.

4.4.3.1 Surface reconstruction

The phenomenon of surface reconstruction refers to the process by which atoms at the surface of a crystal form a different structure than that of the bulk. The free dangling bonds of atoms on the surface are rearranged in order to assume a minimum surface energy. Because of such atom movements the symmetry of the surface changes with respect to the bulk leading to a 2D equilibrium surface structure. Thereby, the atomic distances along the surface become periodically differing and, usually, also the neighboring underlying planes are somewhat affected by strains. Such frequent changes in surface symmetry can be easily detected by surface diffraction techniques such as low-energy electron diffraction (LEED) and reflection high-energy electron diffraction (RHEED) as well as by scanning tunneling microscopy (STM).

In the case of many semiconductors, the simple reconstructions can be explained in terms of a "surface healing" process in which the coordinative unsaturation of the surface atoms is reduced by bond formation between adjacent atoms. Figure 4.11 demonstrates exemplarily the surface reconstruction at the crystal–vacuum interface of Si (100) and GaN (0001). First, we look at the case of silicon (a,b). The formation of an unconstructed Si (100) (1×1) surface leaves two "dangling bonds" per surface Si atom. Now, the relatively small coordinated movement of the surface atoms can reduce this unsatisfied coordination by coming together of Si atom pairs to form surface Si dimers, leaving only one dangling bond per Si atom. This process leads to a change in the surface periodicity whereupon the period of the surface structure is doubled in one <110> direction giving rise to the (2×1) reconstruction. The two related STM images below illustrate the difference between unconstructed (1×1) and reconstructed (2×1) surface states on atomistic scale. Dimer rows along the [110] direction are formed in the course of reconstruction (b). The (100) surface of materials with zinc blende structure behaves almost identically. For instance, an As-terminated GaAs surface forms As-dimers that reduce also the number of dangling bonds and, thus, the free surface energy.

The next example in Fig. 4.11c and d shows the relatively seldom but interesting (4×4) reconstruction of the Ga-terminated (0001) surface of GaN having wurtzite structure (d). During the MBE process it was observed that a rhombus consisting of eight Ga atoms is formed via the (2×2) intermediate state at 500 °C. The related STM images show that at (2×2) reconstruction each rhombus side overlay four (1×1) unit surface cells from non-reconstructed case (c).

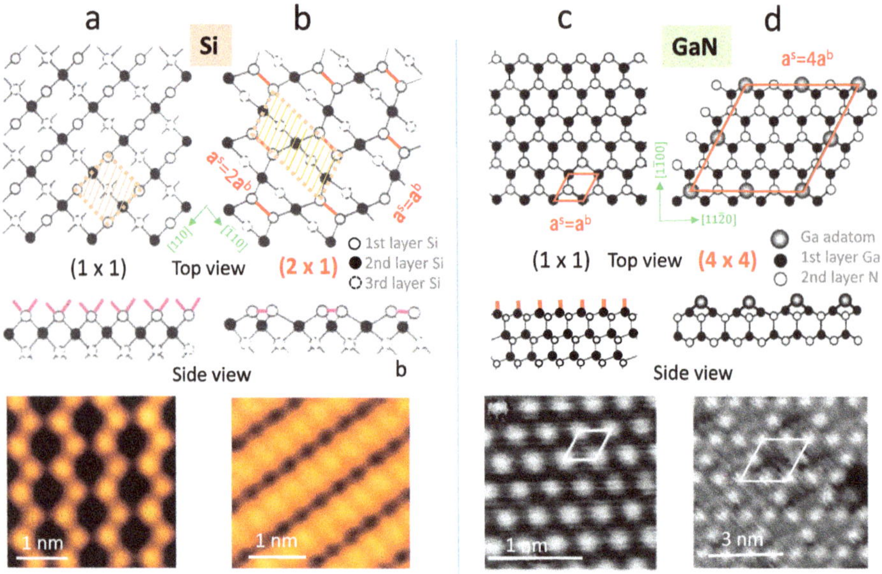

STM topographic image of Si (100) 1x1 surface
Dr. Yokoyama, Yokohama City Univ. 1999

STM of Si (100) 2x1 surface showing dimer rows
A. Crane et al., J. Phys. Conf. Ser. 1079 (2018) 012010

STM of (1x1) (0001)Ga surface of GaN
J. Jia et al., HB Microsc. Nanotech. (Kluwer 2005) p. 55

STM image of (0001) Ga with (4x4) reconstruction
S. Vezian et al., Surf. Sci. 541 (2003) 242–251

Fig. 4.11: Top and side view sketches and STM images of ideal and reconstructed Si (100) and GaN (0001) surfaces (*STM images: with permission of Yokohama City Univ (a), free reuse under license CC BY 3.0 (b), with permission of Springer Nature (c) and Elsevier (d)*).

The surface reconstruction is often given in terms of *Wood's notation* from 1964, wherein the surface crystallography is described by $(hkl)(m \times n)R_\varphi$ with *hkl* the Miller's indices in the bulk, *m,n* the periodicity of the translation vectors of the 2D surface (s) unit cell $\mathbf{a}^s = n\mathbf{a}^b$ and $\mathbf{b}^s = m\mathbf{b}^b$ related to the vectors in the bulk (b), and R_φ the azimuthal rotation angle of the surface unit cell relative to the underlying bulk unit cell. When the simplest 2D unit cell reconstruction without central atom takes place, thus, consisting only of one "repeat unit," the character *p* standing for "primitive" is preceded as $p(m \times n)$. It is common practice to omit *p* and pronounce such structures simply as "*m by n*". On the other hand, when the surface structure differs in that there is an additional atom in the middle/center of the adsorbate unit cell then this is no longer a primitive structure and *c* standing for "centered" is presented as $c(m \times n)$. For instance, a typical reconstructed structure at epitaxial processes on (001) surface of diamond and zinc blende structure proves to be $c(4 \times 4)$. Finally, in case of compounds the *A* or *B* surface cell termination can be expressed by $(m \times n)\alpha$ or $(m \times n)\beta$, respectively.

In this notation, the surface unit cell is given as multiples of the non-reconstructed surface unit cell. For example, the typical calcite (104) (2 × 1) reconstruction means that the unit surface cell is twice as long in direction $\mathbf{a}^s = 2\ \mathbf{a}^b$ and has the same length in direction $\mathbf{b}^s = \mathbf{b}^b$. In Fig. 4.11 these denotations are also added for both demonstrated ex-

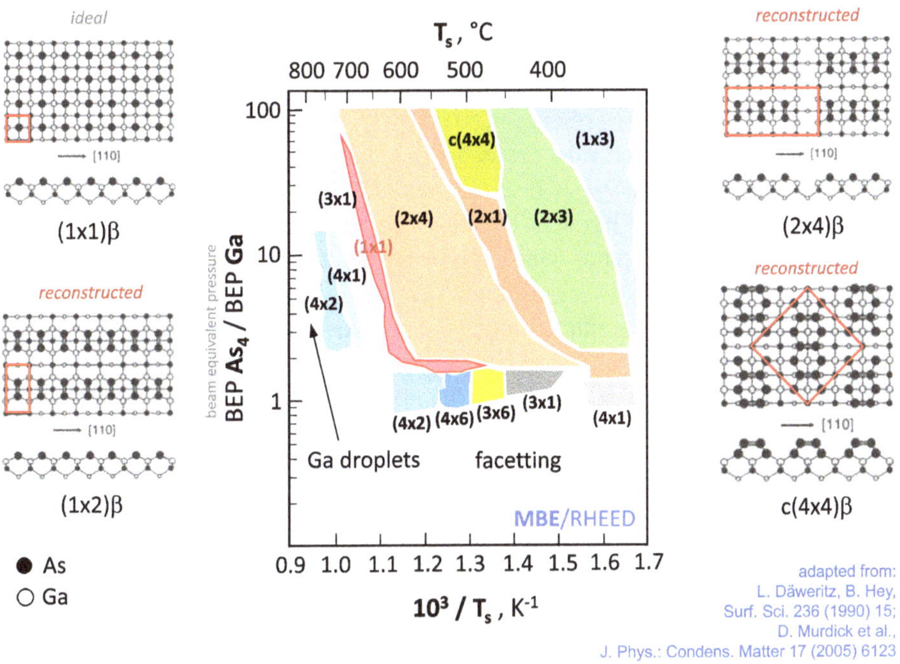

Fig. 4.12: Reconstruction equilibrium structures in the form of p-T surface phase diagram for (001) GaAs performed in a MBE apparatus equipped with a RHEED system and movable ion gauge for beam equivalent pressure (BEP) measurements. Top and side drafts of some reconstructions are added (unit cells are marked in red) (*with permission of Elsevier (diagram) and IOP (top and side sketches)*).

amples of (001) (2 × 1) Si and (0001) (4×4) GaN, respectively. If the unit cell is rotated with respect to the unit cell of the non-reconstructed surface, the factor R_φ is added in degree. For instance, the structure of the Au (001) surface is an interesting example of how a cubic structure can be reconstructed into a distorted hexagonal phase. This hexagonal phase is often referred to as a (28 × 5) structure, distorted and rotated by $R_\varphi \approx 0.81°$ relative to the [011] crystal direction. However, this rotation disappears at $T = 970$ K, above which an unrotated hexagonal structure is observed. Furthermore, at $T > 1170$ K an instantaneously disordered (1×1) structure of square symmetry is recovered.

In general, compound surfaces can develop many different reconstruction patterns depending on the substrate temperature, pressure, pre-transport parameters (vapor association, flux rate, and density) and activities of the constituting atoms along the surface. For instance, under dimer and tetramer antimony flux the surface reconstruction of a GaSb (001) substrate is changing with increasing temperature from (2×5) towards (1×3). Further, at a pressure (flux) ratio p_{As4}:p_{Ga} around 10 the reconstruction periodicity of a GaAs (001) surface translates above 400 °C from (2×3) through (2×1) and (2×4) to (3×1). Such reconstructions are generally equilibrium structures and are presented in the form of *surface phase diagrams* as demonstrated for (001) GaAs in Fig. 4.12. The investigations were performed in a MBE apparatus equipped with a RHEED system and movable ion gauge for beam equivalent pressure (BEP) measurements. As arsen source As_4 was used. It shows the presence of various reconstruction types as a function of temperature T and *BEP* equivalent ratio between As_4 and Ga. Degradation of the surface morphology was observed at $T > 800$ °C and *BEP* ratio < 0.8 by Ga droplet and facet formations, respectively. It has to be mentioned that although not all crystal (substrate) surfaces do reconstruct this phenomenon is extremely frequent and needs to be considered in thin-film depositions very carefully.

At first glance, it's not simple to understand why the reconstruction should reduce the surface free energy, more because the atomic rearrangement entails some strain due to certain elastic deformation (increase) of the surface area. Origin of surface stress could be understood by the nature of *electron density distribution*. Due to the decreased number of nearest neighbors of the surface atoms compared with bulk atoms the local electron density around the atoms near the surface is reduced. As a response the surface atoms are forced to reduce their interatomic distance in order to increase surrounding charge density accompanied by a surface stress, which can be both positive (tensile) or negative (compressive). For instance, semiconductor surfaces forming dimers (see Fig. 4.11b) is the way for it to respond to the tensile stress. If the surface is not clean but comprises several foreign atoms the charge density would then be modified leading to a different surface stress state compared with a perfect clean surface.

Surface stress, first defined in 1878 by *Gibbs*, can be written in simple scalar notation as

$$\sigma^s = \gamma + \frac{\partial \gamma}{\partial \varepsilon} \tag{4.22}$$

with γ the surface-specific free energy and ε the strain. Now it can be easily explained why σ and γ are equal in liquid–gas or liquid–liquid interfaces in which the term $\partial\gamma/\partial\varepsilon$ always equals zero due to the shear nonresistance of γ_L. This is so because if the surface of a liquid is expanded, matter from the bulk rapidly reconstitutes the environment of surface atoms and γ_L remains unchanged. However, $\partial\gamma/\partial\varepsilon$ is not zero in solid surfaces due do the fact that surface atomic structure of solid are modified in elastic deformation. Thus, the derivative $\partial\gamma/\partial\varepsilon$ represents the thermodynamic force on surface atoms to change their location. It's the *driving force for reconstruction* corresponding to the amount of energy gained by structure transformation over the surface stress. Therefore, reconstruction of an ideal surface termination will occur only if the change in $\partial\gamma/\partial\varepsilon$ is large enough to compensate the energy increase due to the energetically unfavorable atom replacement (precisely it has to be noted that differently from the surface free energy γ, which is a scalar, surface stress and strain are second rank tensors σ_{ij}^s and ε_{ij}, respectively. Then, the first term on the right-hand side of eq. (4.22) must by multiplied by the Kronecker symbol δ_{ij})

Usually, a perfect reconstruction feature implies a clean surface in ultrahigh vacuum. In contrast, the adsorption of species onto the surface may enhance, alter, or even reverse the process. The presence of so-called *adsorbates* can modify the surface structure and, thus, the unit cell periodicity. Epitaxial processes, especially the formation of the first nucleation layer and generally the growth kinetics (see lecture part II) of very thin films may be essentially influenced by the reconstruction mode at the used substrate surface. First, the surface diffusion rate and corresponding parameters, like the incorporation diffusion lifetime, can be changed markedly. Further, rapid diffusion that occurs along specific directions of reconstruction enhances the formation of anisotropically shaped islands. Finally, the deposition temperature can be essentially influenced. Thus, in addition to the highest possible preparation quality and purity of an "epi-ready" substrate surface it is imperative to know precisely the epitaxy parameters for control or minimization of the impact of surface reconstruction.

4.4.3.2 Ordering effects in mixed semiconductor thin films

Due to the importance for engineering of new material combinations with extraordinary properties even by epitaxial processes the phenomenon of nonrandom arrangement of binary and ternary alloys, closely correlated with surface reconstruction and reviewed by one of the pioneers *Zunger 1997*, will be presented in Figs. 4.13 and 4.14. Almost all semiconductor devices are based on successive grown thin films made of alternating binary or/and ternary mixed semiconductor alloys $A_{1-x}B_x$ (e.g., $Ge_{1-x}Si_x$) or/and $A_{1-x}B_xC$ (e.g., $Ga_{1-x}In_xAs$, $Ga_{1-x}In_xP$) the electronic properties of which are tailored by varying the composition. In addition, the use of very thin alloys allows the production of special structures such as quantum wells with abrupt changes in bandgap energy. Due to the microscopic scale of the thickness of these layers the energetic surface and interface states become dominant for atom arrangement within the film "bulk." Through the interplay with the effect of surface reconstruction (see Section 4.4.3.1) during epitaxial growth the topmost and few subsurface layers may rearrange in a structure of *atomistic ordering* of the alloy components. Therefore, an *interfacial* effect is taking place. First the discovery of this phenomenon was surprising because for many years it was believed that when two isovalent semiconductors are mixed, they will form a solid solution at high temperature or separated phases at low temperature, but they will never produce ordered atomic arrangements. It was assumed that the mixing enthalpy of an alloy $\Delta h_m(x)$ depends on its global composition x only and not on the microscopic arrangement of atoms. However, careful thermodynamic analysis has shown that certain ordered three-dimensional atomic rearrangements within the top layers minimize the strain energy resulting from the lattice-constant mismatch between the constituents, while random arrangements do not. In particular, there are special ordered structures α that have lower energies than the random alloy of the same composition x, that is

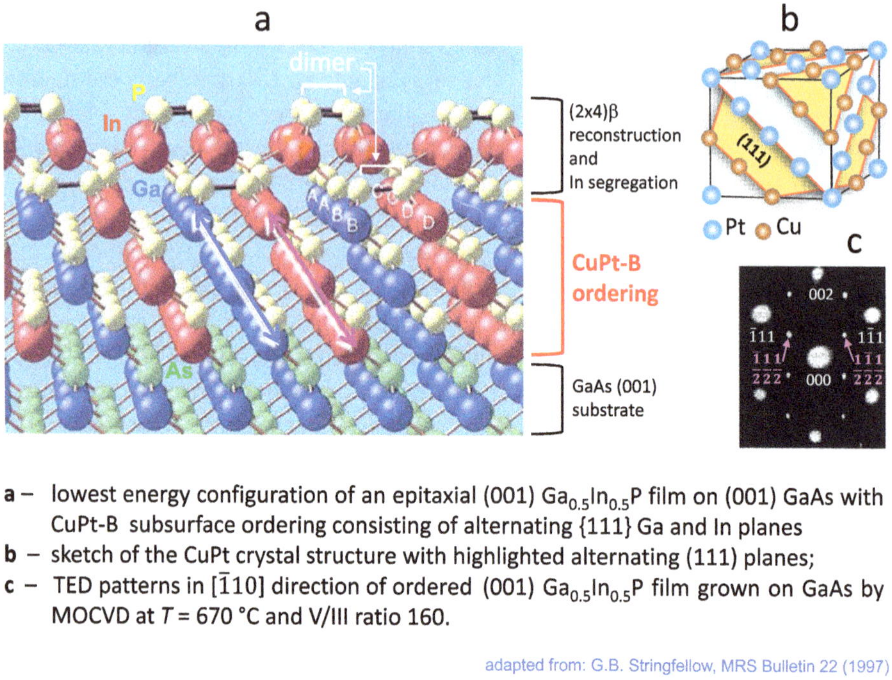
a
b
dimer
P
In
Ga
As
(2x4)β
reconstruction
and
In segregation
CuPt-B
ordering
GaAs (001)
substrate
(111)
Pt
Cu
c
002
000
a – lowest energy configuration of an epitaxial (001) Ga0.5In0.5P film on (001) GaAs with CuPt-B subsurface ordering consisting of alternating {111} Ga and In planes
b – sketch of the CuPt crystal structure with highlighted alternating (111) planes;
c – TED patterns in [1̄10] direction of ordered (001) Ga0.5In0.5P film grown on GaAs by MOCVD at T = 670 °C and V/III ratio 160.
adapted from: G.B. Stringfellow, MRS Bulletin 22 (1997)

Fig. 4.13: Model of an epitaxial $A_{1-x}B_xC$ ($Ga_{0.5}In_{0.5}$ P) film (a) microstructural ordered by CuPt type (b) proved by transmission electron diffraction analysis (c) (*with permission of MRS (a) and Springer Nature (c)*).

$$\Delta h_{\alpha}^{\text{ordered}} < \Delta h_m(x)^{\text{random}} \tag{4.23}$$

where in case of a binary mixed system stands

$$\Delta h_{\alpha}{}^{\text{ordered}} = E_{\alpha}{}^{\text{ordered}} - x\ E_A - (1-x)\ E_B \tag{4.24}$$

and

$$\Delta h_m(x)^{\text{random}} = E(x)^{\text{random}} - x\ E_A - (1-x)\ E_{\text{B}} \tag{4.25}$$

with $E_{\alpha}{}^{\text{ordered}}$ the total energy of a given arrangement α of A and B atoms on a lattice with N sites, E_A and E_B the total energies of the constituent solids, and $x = N_B/N$ the mole fraction. Thus, ordered and disordered configurations at the same composition x can have a different excess enthalpy $\Delta h(x)^{\text{exc}}$ leading to a thermodynamically stable rearrangement in the subsurface layers in contrast to the thermodynamically stable arrangement in an infinite bulk solid.

In addition to CuAu and chalcopyrite basis forms the most frequently observed ordering type for III–V mixed alloys grown epitaxially on (001) oriented substrates is the rhombohedral CuPt structure (Fig. 4.13b), with ordering on one or two of the set of four {111} planes. It occurs when the cation planes take an alternate sequence of A-

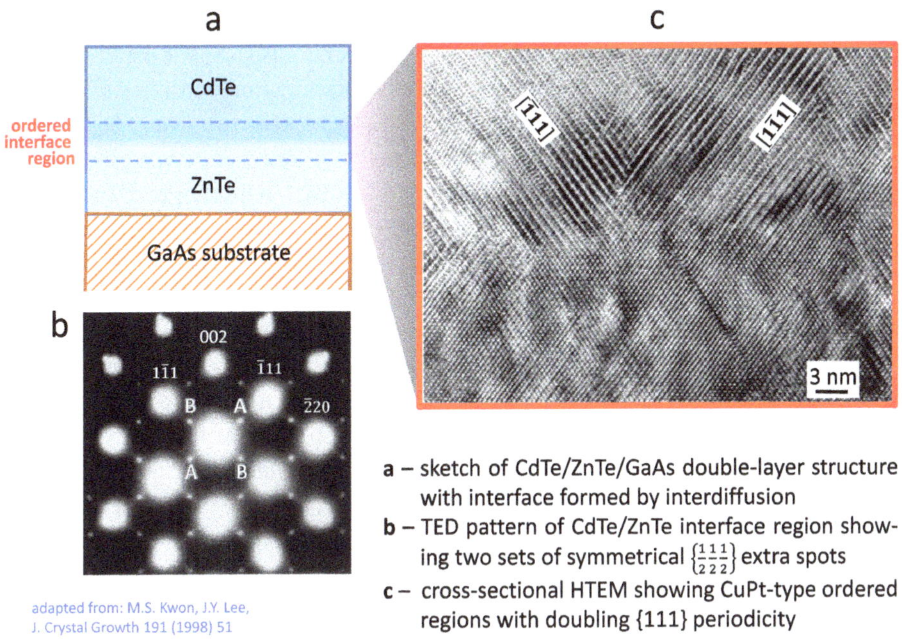
a
c
CdTe
ordered interface region
ZnTe
GaAs substrate
b
002
$1\bar{1}1$
$\bar{1}11$
$\bar{2}20$
B
A
A
B
$[\bar{1}11]$
$[1\bar{1}1]$
3 nm
adapted from: M.S. Kwon, J.Y. Lee, J. Crystal Growth 191 (1998) 51
a – sketch of CdTe/ZnTe/GaAs double-layer structure with interface formed by interdiffusion
b – TED pattern of CdTe/ZnTe interface region showing two sets of symmetrical $\{\frac{1}{2}\frac{1}{2}\frac{1}{2}\}$ extra spots
c – cross-sectional HTEM showing CuPt-type ordered regions with doubling {111} periodicity

Fig. 4.14: Formation of CuPt-type ordered region at heteroepitaxial CdTe/ZnTe film interface obtained by MOCVD on GaAs substrate (a) proved by TED (b) and HTEM (c) (*with permission of Elsevier (b, c)*).

rich and B-rich planes following $A_xB_{1-x}C$ (Fig. 4.13a). The result is usually a superlattice-like structure along $[1\bar{1}1]$ or $[\bar{1}11]$ called (111) B plane, or along $[\bar{1}\bar{1}1]$ or $[11\bar{1}]$ called (111) A plane, respectively. The two ordering directions are therefore called CuPt-type A and B ordering due to the similarity with CuPt crystal structure. In its turn the CuPt structure is stable only for the surface reconstruction that forms [110] rows of $[\bar{1}10]$ group-V dimers on the (001) surface. Figure 4.13a shows the lowest energy configuration of an epitaxial (001) film of $Ga_{0.5}In_{0.5}P$ on a (001) GaAs substrate illustrating the relation between surface reconstruction, surface segregation, and subsurface ordering. When the reconstruction is $(2 \times 4)\beta$ an indium segregation and CuPt-B ordering alternating {111} planes of Ga and In exist. CuPt-B ordering implies that the small cations Ga move under the P dimers (A and B sites) to relax the local compression but the large cations In between dimer rows (C and D sites) to compensate the local tensile stress. Thus, such stress-reducing atomic rearrangement into a new CuPt symmetry of GaP-InP clustering within few subsurfaces below a reconstructed surface minimizes the total energy more effectively than random mixing leading to the thermodynamic phenomenon in eq. (4.23).

Such ordering effects are well detectable by transmission electron diffraction (TED). In $Ga_xIn_{1-x}P$ thin films grown without off-orientation on (001) GaAs substrates additional small superlattice spots appear along the [110] zone axis at $h \pm \frac{1}{2}$, $k \pm \frac{1}{2}$, and $l \pm \frac{1}{2}$ where h, k, l are the fundamental zinc blende reflections (Fig. 4.13c).

The thermodynamically driven effect of microstructural ordering has been also observed in ternary II–VI thin films such as $Hg_{1-x}Cd_xTe$ and $ZnSe_{1-x}Te_x$. It became apparent that such phenomenon of substructuring takes place not only in films of ternary mixed systems but can be obtained at *heterointerfaces between two binary compounds* caused by intermixing of constituent elements too. For instance, CuPt-type ordered (Cd, Zn)Te structures were found at the interface region between CdTe and ZnTe films grown by MOVPE on (001) GaAs substrate at 400 °C. Figure 4.14 shows the scheme of a CdTe/ZnTe/GaAs double-layer structure (a), TED pattern of the interface region through [110] projection with extra spots typically for CuPt-type ordering (b), and a cross-sectional high-resolution TEM image of the interface region showing doubling periodicity in contrast of {111} lattice planes (c). In the TED image A and B represent the possible two superstructure orientations along $[\bar{1}11]$ and $[1\bar{1}1]$ directions, respectively. Their equal probability is obvious from Fig. 4.14c showing electron diffraction doubling in periodicity of $(\bar{1}11)$ and $(1\bar{1}1)$ lattice fringes. Such interfacial effect needs the intermixing of group II elements and rearrangements of atoms by relative high diffusivity being typically for II-VIs. Then an ordering effective sublattice relaxation takes place that reduces the biaxial compressive stress due to the large lattice misfit and the marked difference of thermal expansion coefficients between CdTe and ZnTe. In the end, the excess enthalpy is more effectively reduced than in the case of a disordered atomic arrangement.

Ordering effects have important practical consequences. An interesting order-induced property of thin film alloys is the *reduction of the bandgap energy* compared to that of bulk alloys of the same composition, thus, moving the wavelength further into the infrared. Hence, ordered III–V structures may become useful for IR detectors. But, for that, the materials engineer must be able to control ordering over the entire surface of a wafer in order to replace the spontaneous ordering by a precisely controlled one. Two control measures are the epitaxy temperature and the partial pressure ratio of the constituents within the gas phase. The degree of ordering is possible to determine by in situ surface photoabsorption (SPA), indicating the nature of chemical bonding at the surface.

Recently, it was discovered that the use of *surfactants* may markedly enhance the control of ordering too. Surfactants are elements that accumulate at the surface during growth. For example, the use of isoelectronic group V elements, for example Sb, allows to control the surface structure and, hence, the degree of order and bandgap energy of $Ga_{1-x}In_xP$. Surface photo absorption data indicate that the effect is due to a change in the surface reconstruction, i.e. in the displacement of some surface P dimers by larger Sb dimers.

4.4.3.3 Surface patterning by self-assembling

One of the most investigated phenomena during the last decades is the "self-organized" surface texturing on nanoscale, occurring at the epitaxy of a mixed crystal system onto a substrate with markedly differing lattice parameter. The specific inter-

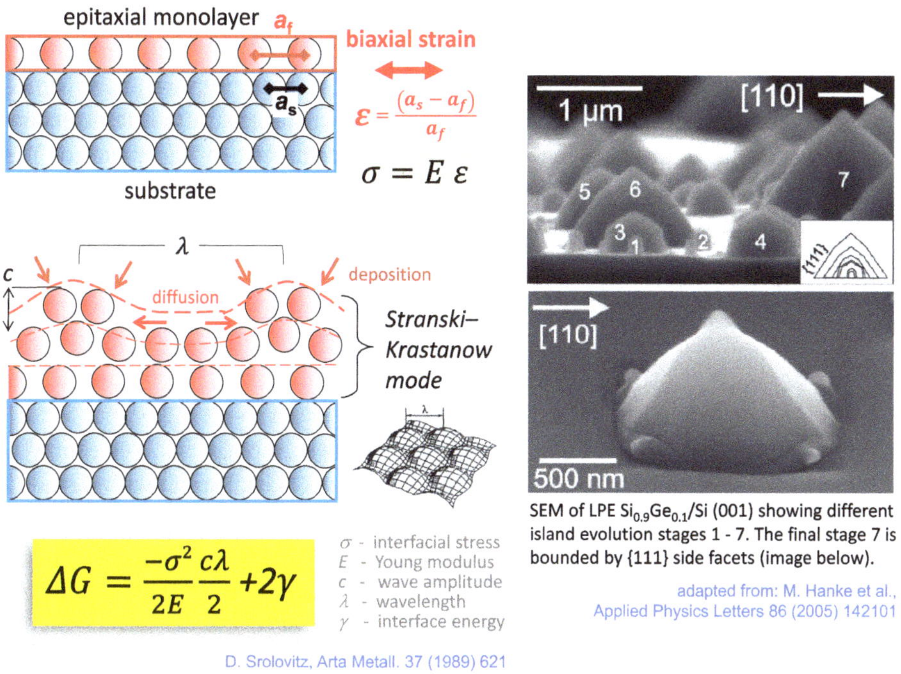
epitaxial monolayer a_f
biaxial strain
a_s
$\varepsilon = \frac{(a_s - a_f)}{a_f}$
$\sigma = E\,\varepsilon$
substrate
λ
c
deposition
diffusion
Stranski–Krastanow mode
λ
$\Delta G = \frac{-\sigma^2}{2E}\frac{c\lambda}{2} + 2\gamma$
σ - interfacial stress
E - Young modulus
c - wave amplitude
λ - wavelength
γ - interface energy
D. Srolovitz, Arta Metall. 37 (1989) 621
1 µm
[110]
5
6
7
3
1
2
4
{111}
[110]
500 nm
SEM of LPE $Si_{0.9}Ge_{0.1}$/Si (001) showing different island evolution stages 1 - 7. The final stage 7 is bounded by {111} side facets (image below).
adapted from: M. Hanke et al., Applied Physics Letters 86 (2005) 142101

Fig. 4.15: Sketch and SEM of a heteroepitaxial LPE $Si_{0.9}Ge_{0.1}$ layer on (001) Si substrate showing the self-organization of a waved film surface, followed by island patterning due to balancing between elastic strain in the film and increase of surface energy (*with permission of Elsevier (equation) and IOP Publ. (images)*).

est is directed on the observed formation of periodically arranged nanospots acting as *quantum dots*. Such patterns offer interesting prospects for the development of new electronic and optoelectronic devices. In particular, if the size, shape, and positioning of those structures can be controlled, they become very attractive for photonic applications such as affecting electromagnetic wave propagation or tunable single-photon light sources. Well known is $Si_{1-x}Ge_x$ as a prototypical system of self-organization of nanostructures during heteroepitaxy. Despite the 4.18% lattice mismatch between Si and Ge, it is possible to grow $Si_{1-x}Ge_x$ pseudomorphically on Si. Such a depositing layer can undergo a transition from planar two-dimensional growth at small thickness to a three-dimensional island structure at higher coverage. The development of a three-dimensional morphology proves to be an alternative to the generation of misfit dislocations as a means to minimize the energy of the heterosystem. This phenomenon is sketched in Fig. 4.15 on the left side.

The origin of self-organized surface patterning is caused by the interplay of interface thermodynamics and growth kinetics (part II of lecture). In this context, an epitaxial system of dissimilar materials can minimize the energy by evolving specific growth morphology such as the *Stranski–Krastanov mode* (see Section 5.2.2), whereupon, initially a two-dimensional layer-by-layer growth is proceeded to wet the surface (Fig. 4.15, left above) but then, as the surface free energy is decisional acting, a transition to three-dimensional morphology takes place. For a mixed crystal film on a heterogeneous substrate under compressive stress, undulation of the surface allows lattice planes to relax toward ripple peaks (Fig. 4.15, left below). This lowers the elastic energy stored in the film, but increases the surface energy as compared to a planar surface. The balance between the reduced stress and increased surface energy defines a critical minimum wavelength λ_c for stable cycloid surface undulations under plane strain conditions. This critical distance was derived by *Chiu and Gao* (1993) based on elastic solutions for a two-dimensional cycloid surface as

$$\lambda_c = \frac{(1-\nu)\pi\gamma}{2\mu(1+\nu)^2\varepsilon^2} \tag{4.26}$$

where ν is the Poisson's ratio of the film material, γ the surface-specific free energy, μ the shear modulus, and $\varepsilon = (a_s - a_f)/a_f$ the misfit-induced biaxial strain with a_s and a_f the lattice constant of substrate (s) and film (f), respectively (see also Fig. 4.15, left above). Surface undulations with wavelength larger than λ_c can stable form via surface diffusion to minimize the system energy (Fig. 4.15, left below). Conversely, for wavelengths smaller than λ_c, it is energetically favorable to reduce the surface energy by filling the troughs so that smoothening is expected. In the case of a Si_{1-x} Ge_x film on a silicon substrate, the Ge atoms will migrate at the crest of the undulations, where the lattice constant is closer to that of bulk unstrained $Si_{1-x}Ge_x$ material. Setting related values for $x = 0.5$ in eq. (4.26) yields λ_c of the order of 100 nm.

Already in 1989, *Srolovitz* estimated the change of energy at lateral stressing in going from a flat to a simplified square-waved surface profile, with wavelength λ and amplitude c. Using this consideration in a first approximation for a sine-like waved film surface (Fig. 4.15, left), the resultant "gain" of surface related energy part is roughly,

$$\Delta G_S = \frac{-\sigma^2}{2E}\frac{c\lambda}{2} + 2\gamma \tag{4.27}$$

with $\sigma = E\varepsilon$ the interfacial stress, and E the Young's modulus [all another terms are specified under eq. (4.26)]. Finally, the wave amplitudes are formed as separate pyramids. Equation (4.27) shows that the formation of a "rough" surface profile lowers the energy of the system, provided that the wavelength $\lambda > 8\gamma E/\sigma^2$. In this way, such a process of "self-assembling" is related to a stress-induced morphological instability (surface undulation), which tends to roughen the film surface and develop cusp-like stress singularities so that the strain energy stored in the film can be effectively released. Thus, again the surface-specific free energy turns out to be crucial. The right side of Fig. 4.15 shows characteristic SEM images of $Si_{0.9}Ge_{0.1}$ pyramidal islands, which are

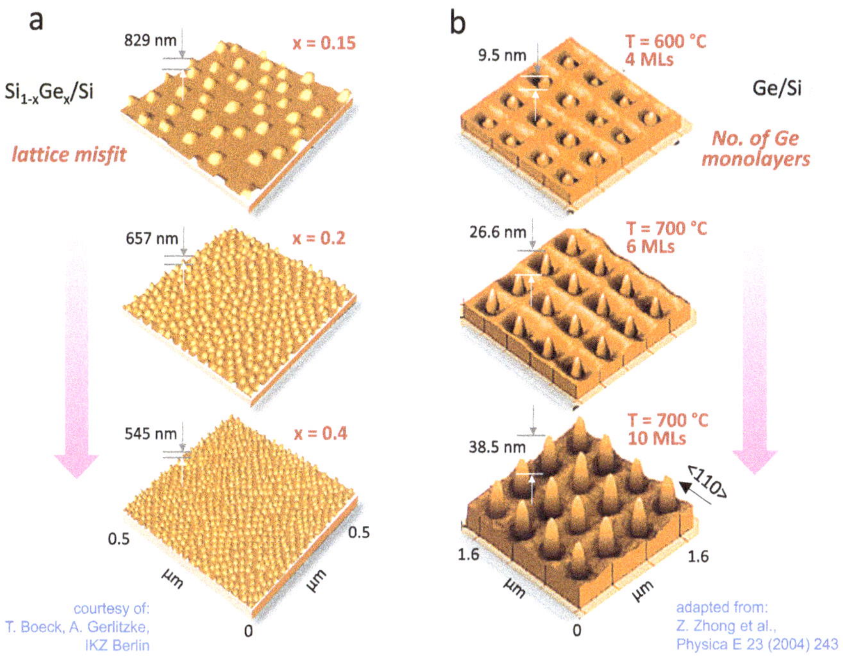

Fig. 4.16: AFM images of $Si_{1-x}Ge_x$ and Ge pyramidal nano-dots grown by LPE on (001) Si substrates without substrate structuring (a) and by MBE on textured substrates (b) in dependence on lattice misfit and number of monolayers, respectively (*with permission of Elsevier (b)*).

grown up from a undulated film surface according to the Stranski-Krastanov mode by applying LPE. The final stage of the pyramids is bounded by {111} facets of minimum free surface energy.

Meanwhile, this process has been studied at vapor and liquid phase epitaxy (ALE, MBE, MOCVD, and LPE) of numerous III–V and II–VI single components, compounds, and mixed alloys. Even arrays of self-assembled *nanowires* have been successfully obtained, especially when atomically stepped ("vicinal") substrate surfaces are used. Figure 4.16 shows AFM images of $Si_{1-x}Ge_x$ and Ge nanopyramids on (001) Si substrates grown by LPE without substrate structuring (a) and by MBE on textured substrates, so-called templates (b). As can be seen from Fig. 4.16a, the size and density of the nanodots can be controlled by the mole fraction x. The higher the related lattice misfit, i.e., the higher the Ge content, the smaller the size and density of the pyramids.

Figure 4.16b shows the results of a seminal technique to obtain well-ordered nanocrystal arrangements. Patterns of Ge islands were grown at quasi highest lattice misfit on (001) Si substrates with lithographically defined two-dimensionally periodic pits. It was observed that the growth rate of the islands on patterned substrates is in general larger than that on flat substrates under identical deposition parameters due to the fact that Ge atoms deposited at the sidewalls of the pits can migrate downwards to the bottom of each island. Further, the island formation on the patterned substrates occurs at an earlier stage than on flat substrates. That means a smaller number of initiating Ge monolayers (MLs) is required for the transition from 2D to 3D growth. TEM investigations revealed that the grown nanocrystals are free of dislocations. Furthermore, the elastic interaction among the islands is also significantly reduced due to their smaller number density in a regular 2D arrangement.

5 Deviation from equilibrium

5.1 Driving force of crystallization

The precondition for crystallization of a stable solid phase within a metastable fluid phase is the *deviation from the thermodynamic equilibrium*. Therefore, to form a crystal, the nutrient (starting, mother) phase (melt, gas, solution or solid) must be in a *metastable state* (in the following we discuss mainly fluid–solid (F-S) phase transitions, although the principles are applicable to the rarer used solid–solid transitions too; see Section 5.2.4). Thus, the Gibbs free energy of a fluid starting phase must exceed that of the crystalline one. Equating the *partial* or *molar Gibbs free potential* g_i of each involved phase i to the related *chemical potential* μ_i [see eq. (3.11)] the difference of chemical potentials at given temperature$T$ and pressure p is

$$\mu_F - \mu_S = \Delta\mu \tag{5.1}$$

referred to as *driving force of crystallization* or *growth affinity*. This holds true not only for the cases of nucleation of a solid phase within a fluid one but also for the propagating fluid–solid interface of each crystallizing bulk or thin-film system. In the following the relations between chemical potential difference and the experimentally used measures for this driving force will be given.

Figure 5.1 demonstrates the required nonequilibrium situation ahead of the crystallization front at crystal growth from melt, solution, and vapor (the sketched examples represent uniaxial crystallization by HB, growth by aqueous solution, and thin-film deposition by MBE, respectively). From the related scheme of a $\mu_i(T)$ phase projection it follows that a certain difference between the equilibrium temperature T_e (e.g., melting point T_m) and the actual temperature at the fluid–solid interface $T = T_g$ (growth temperature) is required to induce the incorporation of building blocks (atoms, molecules) into the crystalline phase (note, in case of an exact thermodynamic equilibrium the absorption transfer of atoms from fluid to solid would to be identical with their back desorption from solid to fluid and therefore no any progressing crystallization could take place). For the case of melt growth the difference $T_e - T$ represents the degree of *undercooling* (*supercooling*) $\Delta T = T_m - T$ acting as related driving force for crystallization. The direct proportionality between ΔT and difference of chemical potentials $\Delta\mu$, derived in *Spec box 5.1*, yields for the liquid–solid ($L \rightarrow S$) phase transition

$$\Delta\mu_{L\rightarrow S} = \Delta h\ (\Delta T / T_m) \tag{5.2}$$

Quite an identical situation exists at growth from solution and vapor (see Fig. 5.1). However, according the thermodynamic parameter specifics instead of the temperature difference the driving forces are expressed by the differences between the actual and equilibrium *concentration* ΔC and *pressure* Δp, respectively [of course, also at these growth methods the temperature remains a decisional operative parameter by

https://doi.org/10.1515/9783111711164-005

Melt growth

supercooled melt:

$$\Delta\mu = \Delta s\, \Delta T = (\Delta h/T_m)\, \Delta T$$

$$\rightarrow \Delta\mu_{L\text{-}S} \approx 100 \text{ J mol}^{-1}$$

Solution growth

supersaturated solution:

$$\Delta\mu = RT \ln(1 + \Delta C/C_e)$$

$$\rightarrow \Delta\mu_{Sol\text{-}S} \approx 500 \text{ J mol}^{-1}$$

Vapor growth

supersaturated vapor:

$$\Delta\mu = RT \ln(1 + \Delta p/p_e)$$

$$\rightarrow \Delta\mu_{V\text{-}S} \geq 1000 \text{ J mol}^{-1}$$

T_e – equilibrium temperature, T_g – growth temperature, T_m – melting point, C_e – equilibrium concentration, p_e – equilibrium pressure, k – Boltzmann constant

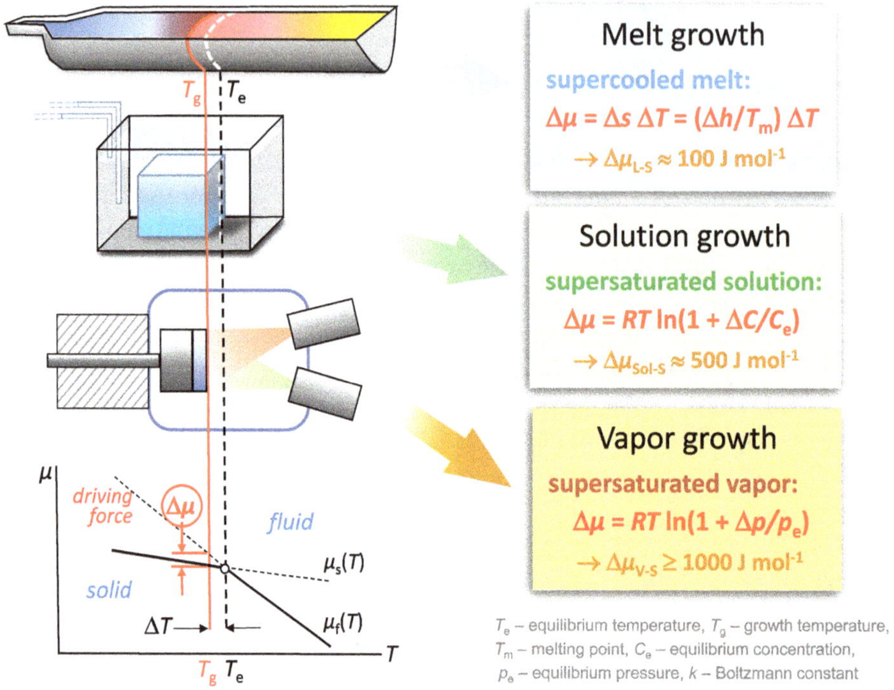

Fig. 5.1: Schemes and comparison of driving force of crystallization at crystal growth processes from the melt, solution, and vapor (for simplification the supercoolings ΔT are sketched equal in all growth principles, which, of course, in reality vary depending on the phase transition).

applying the actual $C(T)$ and $p(T)$ functional dependencies]. That means, the actual concentration C or pressure p of the nutrient fluid phase around the growing crystal must be higher than the equilibrium values p_e and C_e. This difference is designated by the *supersaturation*. Its convenient expressions are

Total supersaturation	$\Delta p = p - p_e$	$\Delta C = C - C_e$	(5.3)
Relative supersaturation	$\Delta p/p_e = (S-1) = \sigma$	$\Delta C/C_e = (S-1) = \sigma$	(5.4)
Partial supersaturation	$p/p_e = S$	$C/C_e = S$	(5.5)
Percentage	$\Delta p/p_e \cdot 100 \ (\%)$	$\Delta C/C_e \cdot 100 \ (\%)$	(5.6)

Clearly, $S < 1$, $S > 1$, and $S = 1$ stand for undersaturated, supersaturated, and saturated vapor or solution, respectively.

As with the melt growth, also the driving forces for crystallization from vapor and solution are quantified by the experimental parameters p and C (see also *Spec box 5.1*).

$$\Delta\mu_{V\to S} = kT\ ln(1 + \Delta p/p_e) \tag{5.7}$$

$$\Delta\mu_{Sol\to S} = kT\ ln(1 + \Delta C/C_e) \tag{5.8}$$

where k is the Boltzmann constant and T the given growth temperature. By using eq. (5.4) for relatively small supersaturations, we get $\Delta\mu/kT = \ln S = \ln(1+\sigma) \approx \sigma \ll 1$.

Spec box 5.1: Driving forces of crystallization

i) Melt–solid transition:
From the comparison of the $\mu_i(T)$ curves in Fig. 5.1, it follows that a stable solid phase requires a difference of the partial chemical potentials between liquid (L) and solid (S) as $\Delta\mu = \mu_L - \mu_S$.The equilibration in molar notation of the Gibbs free energy (see Section 3.2) is

$$\Delta\mu = \Delta g = \Delta h - T\Delta s \tag{B5.1-1}$$

where $T < T_m$ is the actual temperature at the liquid–solid interface, T_m the melting temperature, and Δh, Δs the specific enthalpy and entropy of crystallization, respectively. Inserting the relation $\Delta h = \Delta s\, T_m$ from eq. (2.9) gives the *driving force of crystallization* as

$$\Delta\mu_{L\to S} = \Delta h - T(\Delta h/T_m) = (\Delta h/T_m)(T_m - T) = \Delta h(\Delta T/T_m) \tag{B5.1-2}$$

ii) Vapor–solid transition [*see Markov (2020)*]:
Now the $\mu_i(p)$ curves are relevant. Thus, a stable solid phase does exist when the actual pressure p is higher than the equilibrium pressure p_e, i.e., $p > p_e$. That means the difference of the chemical potential is

$$\Delta\mu = \mu_V(p) - \mu_S(p) \tag{B5.1-3}$$

and more exactly

$$\Delta\mu = [\mu_V(p) - \mu_V(p_e)] - [\mu_S(p) - \mu_S(p_e)] \tag{B5.1-4}$$

where $\mu_V(p_e) = \mu_S(p_e)$ are the partial chemical potentials at the equilibrium. For small deviations from equilibrium eq. (B5.1-4) can be explicitly written as

$$\Delta\mu = \int_{p_e}^{p} \frac{\partial\mu_V}{\partial p}\partial p - \int_{p_e}^{p} \frac{\partial\mu_S}{\partial p}\partial p = \int_{p_e}^{p} (v_V - v_S)\partial p \cong \int_{p_e}^{p} v_V dp \tag{B5.1-5}$$

with the molar partial derivative $(\partial\mu/\partial p) = (\partial g/\partial p) = v = V_V/N_A$ (molar volume). Assuming that the gas is an ideal one ($v_V = kT/p$) after integration of eq. (B5.1-5) and setting $p - p_e = \Delta p$, one obtains the intensified equation (5.7)

$$\Delta\mu_{V\to S} = kT\ \ln(p/p_e) = kT\ \ln(1 + \Delta p/p_e) \tag{B5.1-6}$$

iii) Solution–solid transition:
As was already mentioned under eq. (5.8) for a *rough approximation* the adequate equation (B5.1-6) can be used to express the driving force for solution growth when the solution is treated as an *ideal* one. After replacing C by the mole fraction x, it becomes

$$\Delta\mu_{Sol\to S} = kT\ \ln(x/x_e) = kT\ \ln(1 + \Delta x/x_e) \tag{B5.1-7}$$

Note that solution growth $\Delta\mu$ is often expressed in eq. (B5.1-2) in terms of supercooling ΔT of the solution down to the supersaturation x from the equilibrium temperature T_e at x_e. For this the value $\Delta h_{L\to S}$ must be replaced by the enthalpy of solution $\Delta h_{Sol\to S}$, derived in the *Spec box 2.2*. However, there is no significant difference in the latent heat of melting.

In the predominant cases of *real solutions* the activity a_i must be introduced as it was detailed in Section 3.2.3. Replacing eq. (B5.1-7) by introducing $a = x\,y$ (with y the activity coefficient) we get

$$\Delta\mu_{Sol\to S} = kT\ \ln(a/a_e) \tag{B5.1-8}$$

$$\Delta\mu_{Sol\to S} = kT\ \ln(x/x_e) + kT\ln(\gamma/\gamma_e) \tag{B5.1-9}$$

Using the interaction parameter Ω for a symmetric solution model from eq. (3.42), whereupon $kT\ln\gamma = \Omega(1-x)^2$ at small supersaturation $\ln(1+\Delta x/x_e) \approx \Delta x/x_e$ becomes

$$\Delta\mu_{Sol\to S} \approx kT\left[\Delta x/x_e + \Omega(1-x)^2/\Omega_e(1-x_e)^2\right] \tag{B5.1-10}$$

Sure, for crystallization from solution a favorable solvent B should be found for which the interaction parameter with the matrix A is $\Omega > 1$ in order to obtain a quasi solvent-free crystal [see eq. (3.48)]. Very likely, it can be assumed that Ω/Ω_e differs not so much from unity and the driving force can be calculated based on the ratio of two concentrations x and x_e. However, in more detail this ratio includes the influence of the solvent on the component interaction and growth kinetics under supersaturated and equilibrium conditions, a situation where the driving force needs to be accurately accounted, it being not a trivial research program. Since $\Omega/\Omega_e > 1$ the driving force is slightly enhanced at the growth from solution. Even the process of desolvation before the matrix atoms are incorporated into the growing crystal spends certain activation energy in addition to the supersaturation (see lecture part II).

It is noted that in eqs. (5.7) and (5.8) the product kT is used, which is commonly applied as scale factor for energy in molecular-scale systems. The hitherto used product $RT = kT\,N_A$ is responsible in macroscopic scale, for example, in phenomenological thermodynamic systems. Then the value $\Delta\mu$ is not more intensified and should be, strictly speaking, replaced by ΔG [see eq. (3.11)]. However, to prevent collision with the notation for the system Gibbs energy we will furthermore apply $\Delta\mu$ for the extensive designation of the driving force as is customary in literature too.

The insertion of characteristic growth parameters into eqs. (5.2), (5.7), and (5.8) reveals an approximate comparison between the acting driving forces at the growth from melt, solution, or vapor shown in Fig. 5.1 (of course, the homogeneous nucleation of a new solid phase within a metastable fluid phase without involvement of an artificial seed requires a much larger supercooling and supersaturation than an already presented seed or substrate crystal). According to eq. (5.2) at melt growth the ratio $\Delta h/T_m$ usually falls within the range of 10–150 J/K mol, and values of ΔT are typically 0.1–1 K. Hence, $\Delta\mu_{L\to S}$ becomes not more than about 100 J/mol [note, the undercooling depends sensitively on the atomic nature (roughness or smoothness; see lecture II on kinetics) of the growing interface and can reach at the atomically flat {111} face of dislocation-free silicon crystals $\Delta T \approx 3$–5 K; then after eq. (5.2) with $\Delta h/T_m$ (Si) ≈ 30 J/K/mol the value of $\Delta\mu_{L\to S}$ is 150–200 J/mol that is markedly larger than for defective crystals]. To compare the driving forces of solution and vapor growth with those of melt growth we replace in eqs. (5.7) and (5.8) kT by RT as sketched in Figs. 5.1 and 5.2. Setting in eq. (5.7) $\Delta p/p_e \leq 100$

Epitaxy method	LPE	VPE	MOCVD	MBE
Relative super-saturation σ	~ 0.02 - 0.1	~ 0.5 - 2	~ 50	~ 10 - 100
Driving force $\Delta\mu$ J/mol (T = 1000 K)	~ 165 - 800	~ (3-9) x10^3	~ 3.3 x 10^4	~ (2-4) x10^4

LPE :

$\sigma = \Delta C/C_e = (C/C_e - 1)$
$\Delta\mu = RT \ln(1 + \sigma)$

VPE, MOCVD, MBE :

$\sigma = \Delta p/p_e = (p/p_e - 1)$
$\Delta\mu = RT \ln(1 + \sigma)$

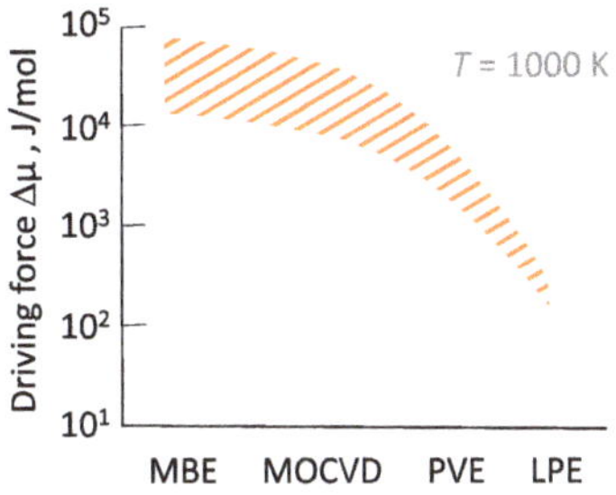

Fig. 5.2: Comparison of driving force at different epitaxy methods.

the value of $\Delta\mu_{V\to S}$ reaches some 1000 J/mol that is around one magnitude of order higher than at melt growth. At solution growth typical values of the relative supersaturation of aqueous solutions $\Delta C/C_e$ are ≤0.1. After eq. (5.8) $\Delta\mu_{Sol\to S}$ becomes around 500 J/mol in the middle between vapor and melt growth. In fact, such a value was found for perfect NaCl crystal growth from aqueous solution. This is also true for the growth from melt–solution like at LPE. Taking an epitaxy temperature of 1000 K and a relative supersaturation of 0.05 the driving force of crystallization becomes ~ 400 J/mol. Thus, the generalized rank order is

$$\Delta\mu_{L\to S} < \Delta\mu_{\text{Sol}\to S} < \Delta\mu_{V\to S} \tag{5.9}$$

Usually, the values of T_e, C_e, and p_e are given by the phase diagrams, which describe equilibria involving pure, unstrained single crystals. However, to describe the growth of strained crystals, such as at heteroepitaxy where the lattice misfit can develop an enormous stress (see Section 4.4.5.2 and lecture part II), terms involving *elastic strain energy* must be added. Similarly, any deviation from the phase diagram conditions (e.g., the application of hydrostatic pressure, electric or magnetic fields) requires minor modifications of the expressions for $\Delta\mu$ too.

It is also important to note that the quasi-conformity of eq. (5.8) with eq. (5.7), where the concentration C replaces the pressure p, proves to be a simplification (being, however, quite applicable for experimental estimations). Strictly speaking, in case of solution growth at least two active components are presented, namely the solute (matrix to be crystallized) A and the solvent B. Thus, it is advisable to replace the concentration C by the mole fraction x, more when intensified thermodynamic parameters are applied, as was introduced in Section 3.2.1. Then, the *activity* as interaction parameter between the components in each phase $a_i = \gamma_i x_i$ (see Section 3.2.1) has to be considered (unfortunately, as it was shown in Section 3.2.3, the finding of the real value of a_i proves to be not an easy task for numerous solution growth systems). The detailed derivations of the driving forces for the three treated phase transitions under consideration with the activity at solution growth are given in the *Spec box 5.1*.

Figure 5.2 compares the approximated relative supersaturation $\sigma = \Delta p/p_e = \Delta C/C_e$ and driving force $\Delta\mu$ for different epitaxy techniques, e.g., LPE, VPE, MOCVD, and MBE. Due to the much higher variability of the vapor pressure selection at growth from the gas phase compared to crystallization from melt–solution the $\Delta\mu$-curve is markedly more spread at MBE and MOCVD than at LPE. Whereas MBE and MOCVD are characterized by the largest driving forces the LPE method proceeds closest to the thermodynamic equilibrium requiring lowest driving force. This is a quite certain consequence for the nucleation processes (see Sections 5.2–5.3) and crystal growth kinetics (lecture part II). As will be seen in the following chapter, for the stage of nucleation of a solid phase within a fluid nutrient without any assisting seed or substrate, largest diving force is required.

5.2 Nucleation

Figure 5.3 shows sketched snapshots of the dynamic process of new phase formation in the fluid phase (vapor, melt, or solution). In a metastable fluid phase of not yet sufficient deviation from equilibrium (case *a*) still random approaches of atoms and molecules and most decay of undercritical clusters are taking place. In other words, close to equilibrium the probability of sufficiently large fluctuations leading to a stable new phase is infinitesimal. This is attributed to the large barrier to phase transition arising from the energy cost for creating an interface between the new and original phase. Only after the driving force $\Delta\mu$ is enhanced by increase of the deviation from equilibrium (i.e., $T_e - T = \Delta T$, $p - p_e = \Delta p$, $C - C_e = \Delta C$ at melt, vapor, and solution growth, respectively) a new stable phase can form (case *b*). In case of a *first-order phase transition* this new phase starts with small complexes of atoms or molecules (nuclei) having a large surface-volume ratio at the beginning. With increasing number of nuclei and their growth the driving force of the system (supercooling, supersaturation) decreases and approaches the phase equilibrium. However, the barrier is ab initio markedly reduced when instead of such mode of homogeneous nucleation a seed or substrate is brought in contact with the metastable fluid phase (case *c*). Then the nucleation and subsequent crystal growth are supported and controlled very effectively. This is the sense of single crystal and thin-film growth. But now let us examine them successively.

Gibbs (1878) and *Thomson (1888)* developed the functional dependence between the size (radius) of the nucleus and difference of the chemical potentials between the starting and nucleating phase [see Section 4.2, eq. (4.10)]. Assuming the simplest case of transition from a supersaturated vapor to a condensed liquid droplet they found that an

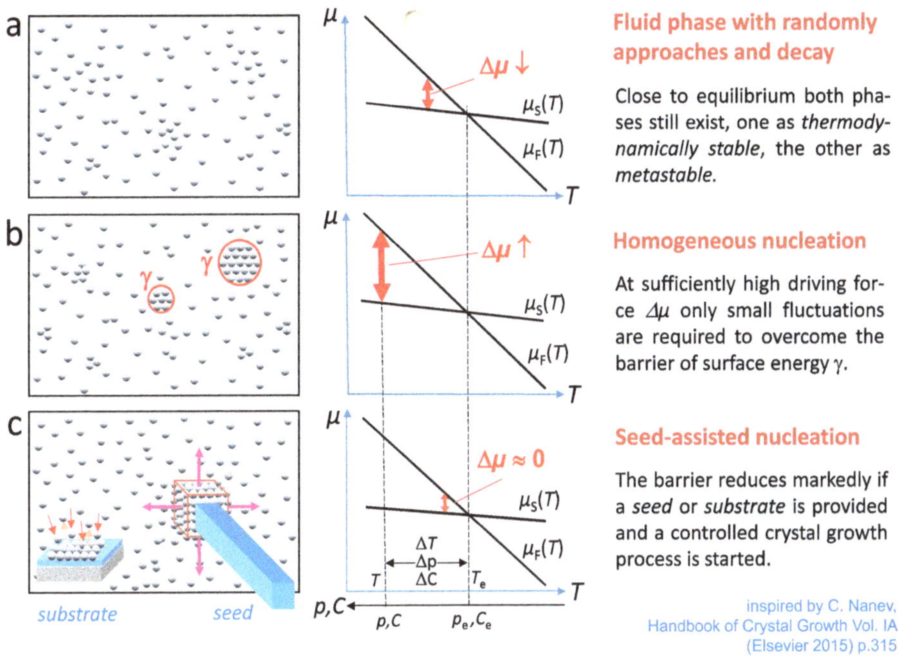

a
b
c
substrate
seed
μ
Δμ ↓
μS(T)
μF(T)
T
Δμ ↑
Δμ ≈ 0
ΔT
Δp
ΔC
Te
p,C
pe,Ce
γ
Fluid phase with randomly approaches and decay
Close to equilibrium both phases still exist, one as thermodynamically stable, the other as metastable.
Homogeneous nucleation
At sufficiently high driving force Δμ only small fluctuations are required to overcome the barrier of surface energy γ.
Seed-assisted nucleation
The barrier reduces markedly if a seed or substrate is provided and a controlled crystal growth process is started.
inspired by C. Nanev, Handbook of Crystal Growth Vol. IA (Elsevier 2015) p.315

Fig. 5.3: Metastable fluid phases and stable new phase by homogeneous and seed-assisted nucleation (*with permission of Elsevier*).

enormous driving force $\Delta\mu_{V\to L}$ is required to overcome the *surface energy effect* hindering the outgrowth of nuclei that are too small. It raises the question of how is it even possible that stable nuclei are formed when their highly convex shape tends to dissolve into the ambient mother phase rather than continue to growth? Already in 1878 *Gibbs* took into consideration that repeated density *fluctuations* in the starting phase could be responsible for the formation of stable nuclei. But this pioneering idea has proved to be insufficient because big nuclei would require unrealistic large fluctuations. The inconsistency was adjusted by the *classical thermodynamic nucleation theory* of *Volmer* and *Weber* in 1926 according to which the main prerequisite for the occurrence of a nucleation process is a sufficiently *high supersaturation (supercooling)*, when even small fluctuations are adequate for the appearance of growth-capable nuclei. In total, very close to equilibrium the probability of sufficiently large fluctuations leading to a stable new phase is infinitesimal. On the other hand, for the occurrence of a nucleation process the establishment of a sufficiently high driving force $\Delta\mu$ is required when only small fluctuations are necessary for critical nuclei formation.

In 1922, it was *Volmer* who introduced the adsorption of growth units onto the crystal surface, their diffusion along the surface, and the generation of two-dimensional nuclei. This was the birth of the quasi-heterogeneous nucleation characterized by a reduced driving force due to the energetic benefit of wetting. Later this led to technical implementations in bulk growth with artificial seeding and epitaxial processes on certain substrates, sketched in Fig. 5.3 c.

Figure 5.4 summarizes examples of nucleation modes currently applied in the bulk crystal growth production. A quasi-homogeneous nucleation regime is used at *industrial crystallization* for production of suspensions comprising of small crystallites such as sugar, table salt, fertilizers, pharmaceutical pills, etc. (Fig. 5.4, left). By pushing a homogeneous multicomponent solution away from thermodynamic equilibrium the driving force creates the plurality of overcritical crystalline nuclei, which after filtration and drying are exposed to the desired particulate product of uniform size. Technically, this process is highly automated and proceeds in giant tanks (crystallizers). Until today, industrial crystallization has to develop into a wide specific research and production branch. For more details there is an extensive related collection of textbooks and publications.

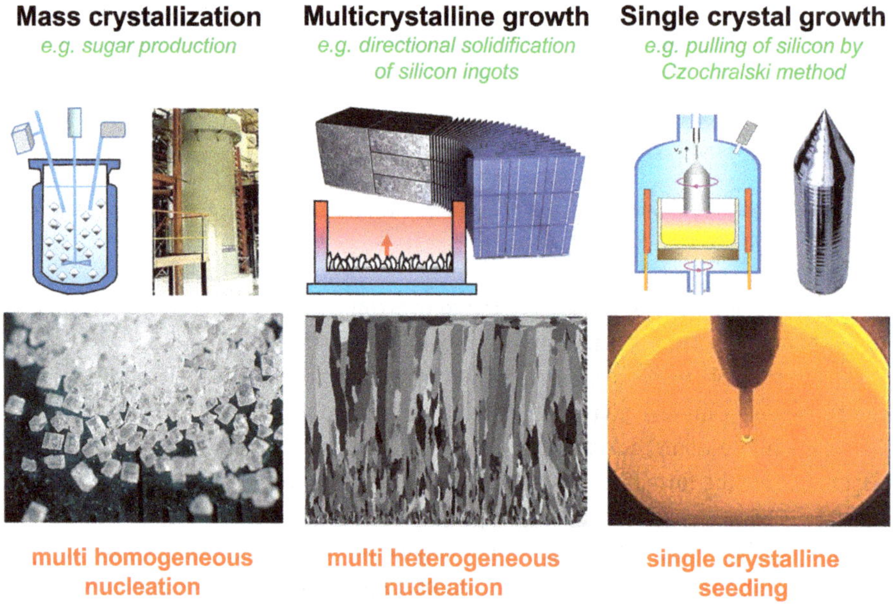

Fig. 5.4: Principles of industrially applied nucleation modes (*images are public domains of Wikipedia*).

Another mode is the directional solidification of products with *multicrystalline structure*, such as cast iron parts, for example. The crystallization starts by multiple heterogeneous nucleation on the cooling bottom of a mold or melt container. As a result, many grains of divergently crystallographic orientations are formed. Depending on the demands the subsequent process is controlled in such a way that the grain size is either kept constant or increasing with the crystallizing ingot height (note even at the beginning of the coalescence the effect of *Ostwald ripening* (1897) is involved supporting the thermodynamic self-selection of size; see Section 5.2.2). One of the most widely used industrial applications today is the directional solidification of multicrystalline silicon ingots, serving for mass production of wafers for solar cells (in the middle of Fig. 5.4). A favorable grain enlargement can be obtained by a convex shape of the crystallization front or positioning of seed panels on the container bottom.

The nucleation barrier is markedly reduced and a single crystalline structure is favored if a crystallization-inducing *seed crystal* is provided (Fig. 5.4, right). Such a mode is widely applied at the bulk growth of single crystals, e.g., by Czochralski, Kyropolos, VB, VGF, FZ, THM techniques.

In the case of diverse epitaxial processes a *matching substrate* that fits both the lattice parameter and coefficient of thermal expansion of the layer to be grown would be the best prerequisite for monocrystalline film depositions. However, in many cases finding such a substrate proves to be difficult or even impossible. Therefore, often the substrate surface must be suitably designed before it is used as base. For instance, this can be achieved by deposition of successively graded buffer layers or artificial surface structuring in order to reduce the lattice misfit or the contact area, respectively.

5.2.1 Homogeneous nucleation

Generally, when providing basic knowledge about the homogeneous nucleation within a native supersaturated (supercooled) phase we should consider the following: it's a spontaneous process generating simultaneously numerous nuclei most of which show uncontrolled divergent crystallographic orientations and, thus, coalesce into a polycrystalline solid structure with many large-angle grain boundaries. Of course, such a feature contradicts the desired monocrystallinity for most devices in electronics, optics, etc. Therefore, these applications require perfect bulk crystals growing predominantly from a single monocrystalline seed. Such step of controlled crystallization requires, above all, the basic knowledge of heterogeneous nucleation shown in Section 5.2.2. Nevertheless, the starting point must be the general principles of homogeneous phase formation, all the more because it is important for mass crystallization and increasingly for the production of nanocrystals.

Until today, the essence of the thermodynamic treatment remains the *classical nucleation theory* (CNT) presented in Section 5.2.1.1 whereupon a microsphere with a critical size forms through fluctuation of atomic configuration, chemical composition, and temperature in the supersaturated (undercooled) liquid. In the CNT, however, the equilibrium crystalline phase is nucleated randomly in a homogeneous completely disordered fluid. This would be immediately transferable to the gas phase. In contrast, recent studies of structural properties of liquids revealed that the supercooled liquid state is no longer homogeneous as assumed by CNT. Thus, crystallization begins with the enhancement of crystal-like bond orientation order followed by translational (density) ordering. In other words, the solid state prefers to nucleate from preordered regions with local orientation symmetry consistent with the crystal. It is obvious that even in the melt of materials with high degree of association due to their marked ionic fraction in the bond energy, such preordering is preferred. Some aspects of the related *non-classical pathway* will be given in Section 5.2.1.2.

5.2.1.1 Classical approach

Let us take the simplified case of formation of a spherical nucleus within a metastable nutrient starting phase comparable with the situation introduced by *Gibbs* in 1876, where a liquid droplet is nucleated within a homogeneous vapor (actually, it has been observed at crystal growth from vapor that initially intermediate liquid nuclei can be formed before they are translated into the solid phase due to the overlapping of the vapor–liquid and liquid–solid metastability regions near the triple point named *Ostwald's step rule*; see below).

The energetic situation is sketched in Fig. 5.5 (left). Concerning the phenomenological thermodynamic treatment we have it to do with two opposite acting forces, namely, (i) the generation of the new stable volume within the metastable starting phase reducing the system enthalpy by energy gain, and (ii) the creation of the associated interface area that consumes energy and, thus, increases the free energy. Sure, even in the moment of generation of very small nuclei the surface share exceeds the volume part considerably. The classical theory of homogenous nucleation provides answers to the question of which nucleus size class belongs to which deviation from equilibrium (supersaturation) and which critical nucleus size is large enough to overcome the barrier (ii) by fluctuation. This is the cardinal point, which equally applies to all modes of phase transition and nucleus shapes.

Therefore, the process of homogeneous nucleation starts spontaneously within the metastable phase after the system attained a sufficient large driving force (supersaturation and supercooling) to overcome the energy-consuming effect of interface formation and to promote the phase equilibrium by outgrowth of nuclei. Due to this competition just-formed nuclei of undercritical size can again disappear. Only nuclei of overcritical size are able to outgrow and contribute to the obtainment of the phase equilibration. Between them an instable situation does exist comparable with the mechanical analog of a ball on a hill. This is inserted in Fig. 5.5 (left) and shows that only a very small push exerted on the ball causes it to roll down, either to the right (which corresponds to growth) or to the left (which corresponds to decay). Even the statistical fluctuations come into play, here. There is always a certain probability that a given number of neighboring atoms or molecules acquire sufficient energy necessary for the formation of stable nucleus.

Assuming the formation of a droplet in the vapor phase, the change of extensive free energy of the system at creation of spherical nuclei is

$$\Delta G = -\Delta G_V + \Delta G_{IF} \tag{5.10}$$

$$\Delta G = -\frac{4\pi r^3}{3\Omega_V}\Delta\mu + 4\pi r^2\gamma$$

γ - free surface energy,
Ω_V - specific particle volume,
$\Delta\mu$ - driving force

Max Volmer
(1885 – 1965)

$$\Delta G^* = \frac{1}{3}4\pi r^{*2}\gamma$$

decay growth

surface part
$\sim +r^2$

resulting $\Delta G(r)$

volume part
$\sim -r^3$

$\Delta G(r)$

$r^*(\Delta T)$

$$\Delta G^* = \frac{16\pi}{3}\frac{\gamma^3\Omega_V^2}{\Delta\mu^2}$$

$$r^* = \frac{2\gamma\Omega_V}{\Delta\mu}$$

$$\Delta G^* = \frac{16\pi}{3}\left(\frac{\Omega_V}{\Delta s\Delta T}\right)^2\gamma^3$$

ΔT – undercooling,
Δs - entropy change

$$r^* = \frac{2\gamma\Omega_V}{\Delta s\Delta T}$$

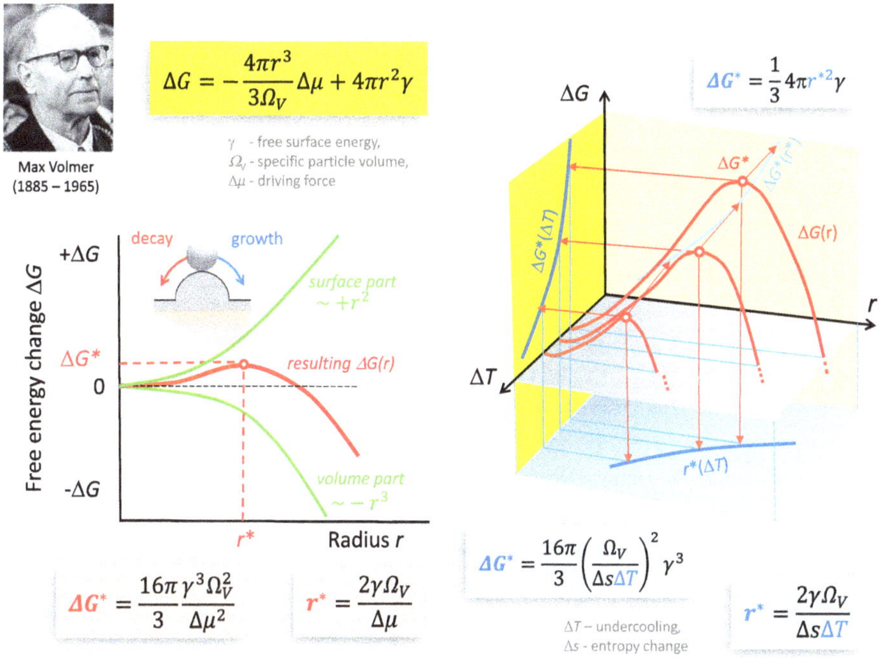

Fig. 5.5: Overview of the general formulas of free energy change versus nucleus radius and related graphical functions of homogeneous nucleation (*the portrait of M. Volmer is public domain*).

where $-\Delta G_V$ associates with the energy gain to form and increase the volume of the stable phase being therefore preceded by a minus sign, and $+\Delta G_{IF}$ stands for the expenditure of energy (plus sign) required for creation and increase of the interface between nucleus and nutrient phase. After insertion of the related parameters for the volume and surface, the energy change is

$$\Delta G = -\frac{4\pi r^3}{3}\Delta\mu/\Omega_V + 4\pi r^2\gamma \tag{5.11}$$

where r is the nucleus radius, Ω_V the specific volume of the building blocks (atoms, molecules), $\Delta\mu$ the driving force (here the difference between the chemical potentials of the vapor and liquid), and γ the free interface energy. Figure 5.5 left shows the summarized $\Delta G(r)$ function of eq. (5.11) in red consisting of the volume and surface term for a given supersaturation $\Delta\mu$. Its maximum at ΔG^* determines the radius of the critical nucleus r^*, which is calculated by the derivation of eq. (5.11) and setting $\partial DG/\partial r = 0$ as

$$\frac{\partial\Delta G}{\partial r} = -\frac{4\pi r^2}{\Omega_V}\Delta\mu + 8\pi r\gamma = 0 \tag{5.12}$$

so that the critical nucleus radius is

$$r^* = \frac{2\gamma\Omega_V}{\Delta\mu} \tag{5.13}$$

Inserting eq. (5.13) into eq. (5.11) gives the maximum in total free energy change, also named *work of nucleation*

$$\Delta G^* = \frac{16\pi}{3}\frac{\gamma^3\Omega_V^2}{\Delta\mu^2} \tag{5.14}$$

which is reached at the critical nucleus size. The physical meaning is that the nucleus can grow only if its actual radius exceeds the critical one ($r > r^*$) because only then it decreases the total free system energy. On the other hand, at $r < r^*$ the nucleus is unstable and dissociates. Note that the "*shape factor*" $16\pi/3$ in eq. (5.14) belongs to a spherical nucleus. In comparison, crystalline nuclei show polyhedral habits with a somewhat enlarged shape factor of about two (see Spec box 5.2).

Substituting the supersaturation $\Delta\mu$ in eq. (5.14) by the potential difference of the *Gibbs–Thomson equation* (4.10) for a critical radius obtains

$$\Delta G^* = \frac{1}{3}4\pi r^{*2}\gamma \tag{5.15}$$

which means that the Gibbs free energy required to form a critical nucleus is equal to one-third of the interface-related energy term in eq. (5.11). This relation is projected in the $\Delta G(r)$ plane within the 3D picture in Fig. 5.5 right, in which solid spherical nuclei within a liquid phase are assumed. In the same projection space it is also demonstrated how the free energy is changed with undercooling when the driving force is expressed in eq. (5.2) as $\Delta\mu = \Delta h(\Delta T/T_m) = \Delta s \Delta T$, where Δs is the change of entropy of fusion. It is obvious that the maxima ΔG^* of the $\Delta G(\Delta T)$ curves are reducing with increasing undercooling and move toward smaller critical nucleus diameter (Fig. 5.5, right).

In case of homogeneous nucleation of a droplet within a vapor, the driving force is determined by the supersaturation. Depending on the vapor element, its consistence, and temperature, the relative supersaturation $S = p/p_e$ is in the range of 2 to 6. For instance, a critical water droplet nucleating within a watery vapor at $T = 275$ K needs a value of $S \approx 4$. Using eq. (5.7) the extensive driving force $\Delta\mu_{V\to L} = kT \ln S$ is $\approx 4.8 \times 10^{-21}$ J and the intensive $RT \ln S \approx 3$ kJ/mol. Then, the critical radius is estimated according to eq. (5.13) after insertion of $\Delta\mu_{V\to L} \approx 4.8 \times 10^{-21}\text{J}$, $\gamma \approx 75 \times 10^{-7}\text{J/cm}^2$, and $\Omega_V = m/(N_A\rho) = \Omega_V = m/(N_A\rho) = 18\,\text{g/mol}/\left(6.023 \times 10^{23}\text{mol}^{-1} \times 1\,\text{g/cm}^3\right) \approx 3 \times 10^{-23}\text{cm}^3$ with m and ρ being the molar mass and density of water, respectively. Thus, at partial supersaturation of $S = 4$ the critical radius of a water droplet within a vapor becomes $r^* \approx 0.9 \times 10^{-7}\text{cm} \approx 1\,\text{nm}$. That means, the volume of the critical water droplet ($4/3\,\pi\, r^{*3}$) consist of about $n^* \approx 130$ water molecules and the maximum of free energy change [eq. (5.15)] yields $\Delta G^* \approx 7\times 10^{-19}$ J. Sure, with decreasing supersaturation the critical droplet radius increases.

Much theoretical estimation by the classical nucleation theory has been done for a long time. However, realistic experimental verification were absent until recently. One of the fascinating experiments was performed by *Zhang* and *Liu* in 2004. They observed homogenous nucleation processes during assembling of 2D colloidal monolayers by controlling an alternating electric field (AEF). As AEF-driven nucleating substance they used polystyrene spheres with diameter 0.99 nm pending in deionized water. Selected images 1–4 are shown in Fig. 5.6. The revealed statistics of decay and growth of subcritical clusters and critical nuclei follow the theory quite well. As can be seen, the nuclei have to be of a critical size before they become thermodynamically stable (yellow circle). In comparison, subcritical nuclei do usually dissolve (red circle). Only seldom they grow by chance. The middle $n(t)$ curve shows that the nucleation characterized by successive agglomeration of the sphere n starts from a nonstationary state and gradually approaches a stationary state of critical number n^* after an initiating time t_i. Then the distribution of nuclei is time-independent. Thus, the critical size of nuclei is definite only at a stationary state. The right curve shows the experimentally detected nucleation rate J standing for the average number of newly formed supernuclei per unit time in a unit area. It depends on the driving force (here supersaturation σ) very sensitively and is of exponential character, which means that at too low driving force no nucleation can occur.

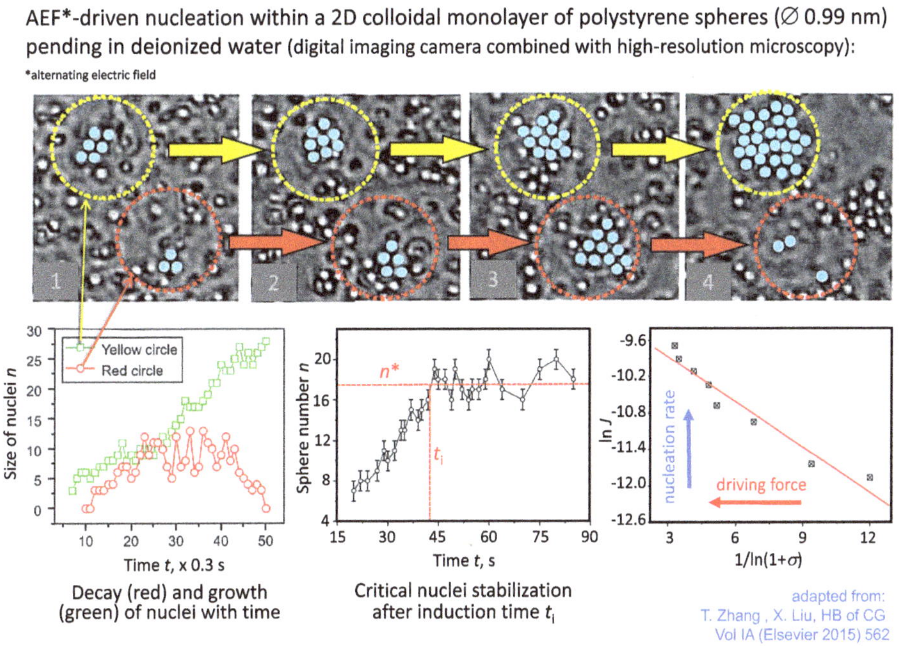

Fig. 5.6: Experimental verification of classical nucleation theory (*with permission of Elsevier*).

Consequently we come to the conclusion that the nucleation process is of statistical nature, which is characterized by possible decay and growth of the presented nuclei with critical size. Therefore, the rate of growth-capable nuclei correlates with the probability to achieve sufficient fluctuation energy. For this reason, in 1926, *Volmer* and *Weber* were the first who expressed the *nucleation rate J* by the Boltzmann factor

$$J = J_0 \; exp\left(\frac{-\Delta g^*}{kT}\right) \tag{5.16}$$

which is the number of nuclei per unit time (s) with J_0 the pre-exponential factor, which later was determined by applying the kinetic principle of a nucleation process running as a series of reaction of collision and afterward absorption or desorption of atoms or molecules.

To express the nucleus size by its number of atoms or molecules we can apply

$$\frac{\text{volume of nucleus}}{\text{volume per atom}} = \frac{\text{volume of nucleus}}{\text{volume of unit cell/number of atoms per unit cell}} \tag{5.17}$$

The probability of finding a critical nucleus consisting of a critical number of atoms or molecules n_i^* by adding eq. (5.14) is

$$n_i^* = n_i \; \exp\left(\frac{-\Delta G^*}{kT}\right) = n_i \; \exp\left(-\frac{16\pi\, \gamma^3 \Omega_V^2}{3kT \; \Delta\mu^2}\right) \tag{5.18}$$

where n_i is the number of single atoms or molecules when the system is in equilibrium. After *Becker* and *Döring* in 1935 determined the kinetic nature of the pre-factor, the stationary nucleation rate by adding eq. (5.17) for homogeneous nucleation is

$$J = Z\; \upsilon^{+*} n_i^* = Z\; \upsilon^{+*} n_i \; \exp\left(\frac{-\Delta G^*}{kT}\right) \equiv \frac{2D}{\lambda^2} \; \exp\left(\frac{-\Delta G^*}{kT}\right) \tag{5.19}$$

with Z being the *Zeldovich factor* $Z = Z_{n^*}/C_{n^*} - Z_{n^*+1}/C_{n^*+1}$ where Z_{n^*} is the steady-state cluster size distribution with C_{n^*} the equilibrium concentration of n-sized clusters, and υ^{+*} the probability of attachment of an atom or molecule at the critical nucleus replaceable by the rate j^+ at which atoms or molecules attach to the nucleus per unit time with same transition probability in each direction. Then the hopping rate is given by $Z\, j^+$, which can be expressed in terms of the mean free path λ and the mean free time τ as $Z\, j^+ = 1/\tau = 2D/\lambda^2$. Consequently, a relation of $Z\, j^+$ in terms of the diffusion coefficient D is obtained. In order to consider the temperature dependence the value of D can specified by the Einstein-Stokes relation for a spherical case as $D = \frac{kT}{6\pi\eta\lambda}$ where η is the dynamic viscosity of liquid at given T. For more details the reader is referred to the extensive literature.

Mostly, the *experimental investigation* of maximum supersaturation or supercooling proves to be hampered by the presence of foreign particles (impurities), which reduces the driving force of homogeneous nucleation by heterogeneous reactions (see

Section 5.2). First *Turnbull* in 1950 concluded that if a liquid is subdivided into very small droplets, most droplets would not contain nucleation-assisting particles and the nucleation should occur in the bulk of the liquid. He succeeded by a first experiment with high-purity mercury and observed that the small droplets could undercool to about 0.8 T_m. Sometime later he summarized identical experiments with diverse metals and found a good accordance whereupon maximum supercooling yields around 20% of the melting temperature in all cases. In Tab. 5.1 Turnbull's and some newer results are tabulated. As can be seen, even highly purified water undercools until the homogeneous nucleation by almost ~18% (425 K).

Tab. 5.1: Degree of maximum supercooling of small droplets of various materials (*with permission of AIP Publ., Elsevier and Springer Nature; the portrait of D. Turnbull is from the free encyclopedia of Wikipedia*).

David Turnbull (1915 - 2007)

Material	Melting point T_m, K	max. Supercooling $T_m - T = \Delta T$, K	Percentage $\Delta T/T \times 100\%$	Reference
Hg	234.3	58	24.7	D. Turnbull, J. Chern. Phys. 18 (1950) 768
Ga	303	76	25	D. Turnbull, J. Metals 188 (1950) 1144
Sn	505.7	105	20.8	B. Vonnegut, J. Colloid Sci. 3 (1948) 563
Bi	544	90	16.6	D. Turnbull et al. J. App. Phys. 21 (1950) 804
Pb	600.7	80	13.3	D. Turnbull, J. Metals 188 (1950) 1144
Al	931.7	130	14	D. Turnbull et al. J. App. Phys. 21 (1950) 804
Ge	1231.7	227	18.4	Δ
Au	1336	230	17.2	Δ
Cu	1356	236	17.4	Δ
Si	1683	320	19	C. Panofen, Mat. Sci. Eng. A 449 (2007) 699
Ni	1725	319	18.5	D. Turnbull et al. J. App. Phys. 21 (1950) 804
Fe	1803	295	16.4	Δ
Pt	2043	377	19	T. Itami et al., Mat. Transact. 51 (2010) 1510
NH_3	195	40.3	20.6	D. Thomas et al. Chem. Soc., 1952, p. 4569
H_2O	273.2	40.5	14.8	Δ
KCl	1045	171	16	E. Buckle, Proc. Roy. Soc. A259 (1961) 325
NaCl	1074	169	16	Δ

Reviews: D. Turnbull, J. Appl. Phys. 21 (1950) 1022;
K. Jackson Ind. Eng. Chem. 57 (1965) 29

Spec box 5.2: The critical solid nucleus of polyhedral shape

At phase transitions from vapor, solution, or melt into a crystalline solid nuclei with polyhedral habit are formed. After *Nanev* (2015) at a homogeneous nucleation of *melt–solid transition*, eq. (5.10) is modified by the shape of the nucleus volume V_{cr} and anisotropy of the interface energy as

$$\Delta G = -\frac{V_{cr}}{\Omega}\Delta\mu + \sum \gamma_{hkl}A_{hkl} = -n\Delta\mu + \sum \gamma_{hkl}A_{hkl} \qquad \text{(B5.2-1)}$$

where $n = V_{cr}/\Omega$ is the number of atoms ore molecules in the nucleus, γ_{hkl} and A_{hkl} the interface energy and surface area of each polyhedron faces specified by the Miller indices *hkl*, as was introduced in eq. (4.15). For instance, when a solid nucleus shows a simple cube shape with {100} faces of edge lengths L the change of free energy according eq. (B5.2-1) is

$$\Delta G = -\frac{L^3}{\Omega}\Delta\mu + 6L^2\gamma_{hkl} \qquad \text{(B5.2-2)}$$

Due to the uniformity of $\gamma_{\{100\}} = \gamma$, the derivative $\partial\Delta G/\partial L = 0$ gives the edge length of a critical cube-shaped nucleus to be

$$L^* = \frac{4\Omega\gamma}{\Delta\mu} \qquad \text{(B5.2-3)}$$

with the potential barrier of nucleation, which is calculated by insertion of L^* into (B5.2-2) as

$$\Delta G^* = 32\frac{\gamma^3\Omega^2}{\Delta\mu^2} \qquad \text{(B5.2-4)}$$

Comparing eqs. (B5.2-4) with eq. (5.14) the *shape factor* is $16\pi/3$ and 32 for a sphere and cube, respectively. With other words, when the interface energy is the same for sphere and cube the barrier of nucleation for a cube is about two times more than for a sphere.

Defining the critical nucleus size by the *number of atoms or molecules* in it $n = V_{cr}/\Omega$ the shape of both volumes have to be considered. Taking the simplest model of a cubic primitive crystal lattice consisting of tiny building units of volume $\Omega = a^3$ with edge length a, the total surface area A of a nucleus of such a shape with edge length $L = n^{1/3}$ is $A = 6\,a^2\,(n^{1/3})^2 = 6\,a^2\,n^{2/3}$ and eq. (B5.2-2) yields

$$\Delta G = -n\Delta\mu + \gamma_{\{100\}}6a^2n^{2/3} \qquad \text{(B5.2-5)}$$

After $\partial\Delta G/\partial n = 0$ the critical number of building units in a cube-shaped critical nucleus is

$$n^* = 64\frac{\gamma^3\Omega^2}{\Delta\mu^3} = 64\frac{\gamma^3 a^6}{\Delta\mu^3} \qquad \text{(B5.2-6)}$$

Considering the melt–solid transition we express the driving force of crystallization by the degree of *supercooling* from eq. (5.2)

$$\Delta\mu_{l\to s} = \Delta h(\Delta T/T_m) \qquad \text{(B5.2-7)}$$

where Δh is the heat of fusion and T_m the melting point. Therefore,

$$n^* = 64\frac{\gamma^3 a^6}{[\Delta h(\Delta T/T)]^3} \qquad \text{(B5.2-8)}$$

Taking the characteristic maximum of relative supercooling of metallic melts $\Delta T/T_m \approx 0.2$ and intensifying the intensive heat of fusion by $\Delta h/N_A = 13\ \text{kJ/mol}/6.023\times10^{23}\text{mol}^{-1} = 2.16\times10^{-20}\text{J}$ (polonium with simple cubic structure) the driving force is $\Delta\mu_{L\to S} \approx 4.3\times10^{-19}\text{J}$. At usual values for the interface energy of metals $\gamma \approx 15\times10^{-6}\text{J/cm}^2$ and edge length $a = 0.33\times10^{-7}\text{cm}$ the critical number of atoms n^* yields ~3483.

A likewise calculation for nuclei of cubic structure in high-purity silicon melt at the highest measured supercooling $\Delta T \approx 300$ K (see Fig. 5.8) gives the number of atoms in a related critical nucleus of about 2000. In comparison, at a much smaller melt supercooling of $\Delta T \approx 3$ K the number of Si atoms in the critical nuclei increases up to about 2×10^9.

Today the experimental analysis of homogeneous supercooling is effectively provided on contactless melt droplets positioned in facilities counteracting the gravity effect by *electrostatic* or *electromagnetic levitation*. Both are demonstrated in Fig. 5.7. Whereas the first method utilizes a laser to melt the sample, in the second one induction heating is applied. In the absence of a crucible, levitated droplets can undergo considerable undercooling, and the nucleation process is free from effects related to contact with refractory walls. This permits investigation of many unique solidification phenomena including degree of supercooling, speed of solidification front, phase selection, and morphology. According the inserts between c and d in Fig. 5.7 the latent heat generated upon solidification raises the temperature of the solidified phase, which is identified as the light area *S*, while the dark area represents the liquid phase *L*.

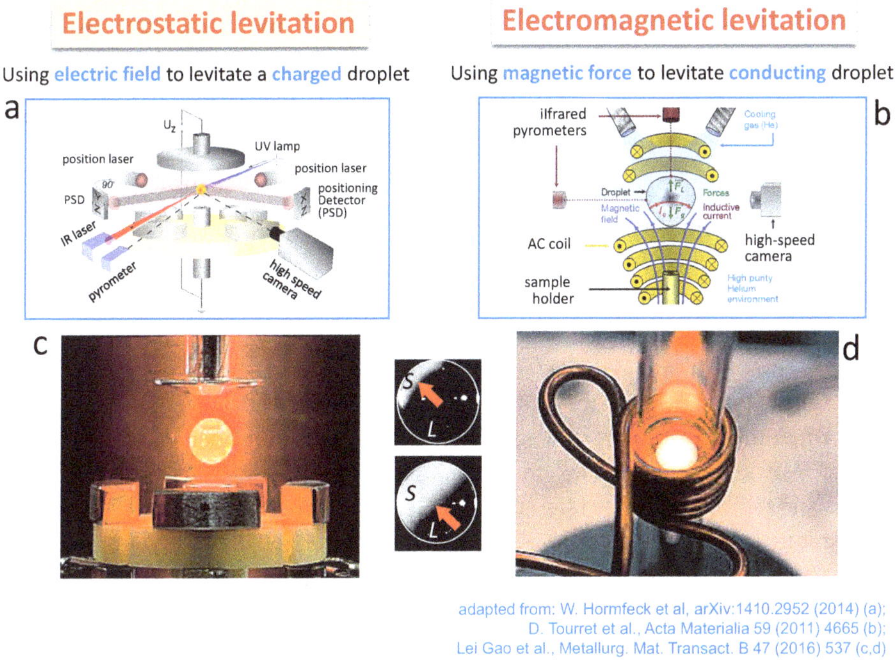

Fig. 5.7: Supercooling analysis on levitating droplets (*with permission from Arxis with CCBY4.0 license (a), Elsevier (b), and Springer Nature (c, d)*).

Using the electromagnetic levitation method the functional dependence between growth velocity and degree of undercooling were detected in silicon and Si–Ge melt droplets by *Panofen* and *Herlach* in 2007. The experimental curve and its comparison with theoretical one is shown in Fig. 5.8a. Supercoolings as high as 300 K have been obtained at both materials accompanied by extremely high growth velocities of more than 15 m/s.

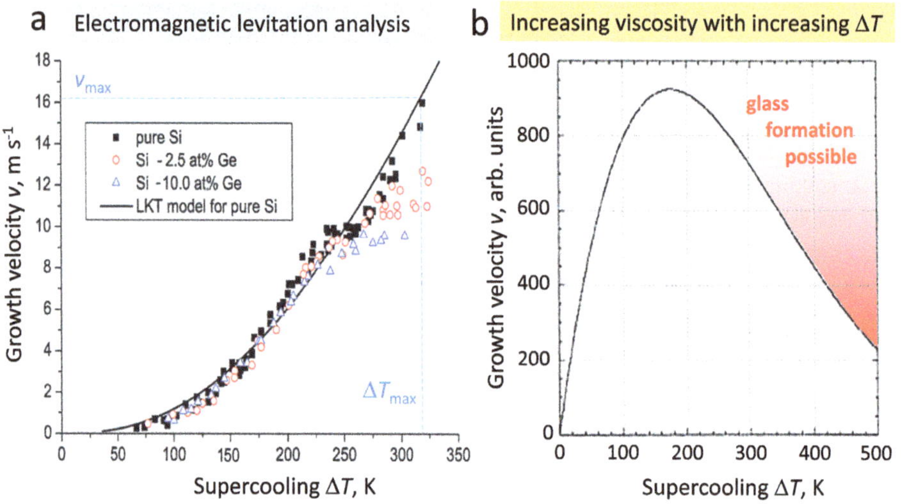

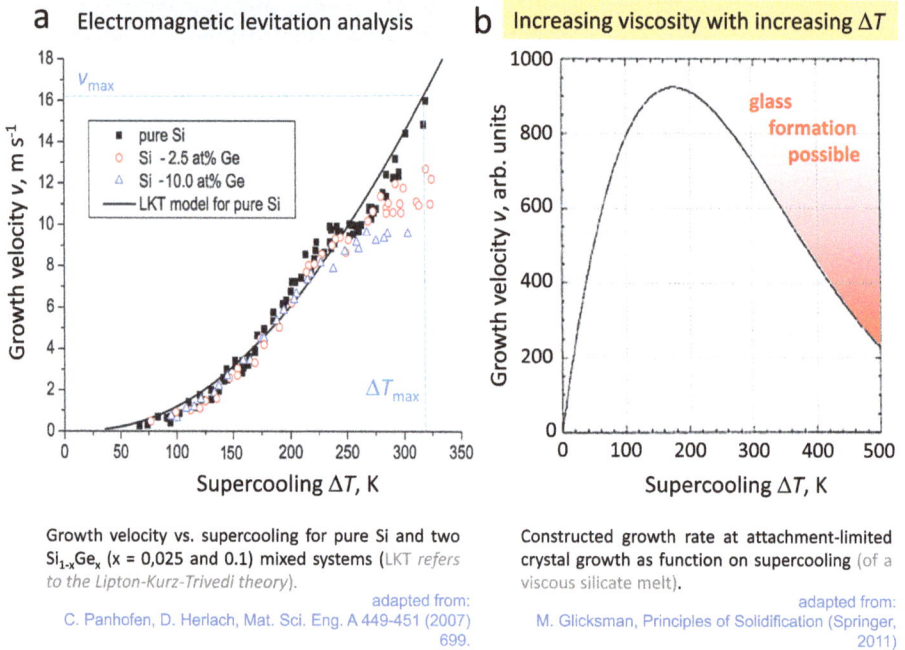

Growth velocity vs. supercooling for pure Si and two $Si_{1-x}Ge_x$ (x = 0,025 and 0.1) mixed systems (LKT *refers to the Lipton-Kurz-Trivedi theory*).

adapted from:
C. Panhofen, D. Herlach, Mat. Sci. Eng. A 449-451 (2007) 699.

Constructed growth rate at attachment-limited crystal growth as function on supercooling (of a viscous silicate melt).

adapted from:
M. Glicksman, Principles of Solidification (Springer, 2011)

Fig. 5.8: Correlation between solidification velocity versus supercooling of levitating Si and Si–Ge droplets (a) and illustration of the influence of the increasing melt viscosity (b) (*with permission of Elsevier (a) and Springer Nature (b)*).

Thus, the moment of nucleation can be markedly depressed by high cooling rates, more when the growth rate of a given crystalline material is restrained due to the kinetic resistance. That means, on the one hand the diffusivity of building blocks (atoms and molecules) toward the still undercritical nuclei is decreasing with temperature reduction due to the exponential increase of the melt viscosity; on the other hand, the possible presence of atomically smooth solid–liquid interface retards the attachment of new atomic layers (see lecture part II). In other words, the moment of recovery of phase equilibrium by homogeneous nucleation will be further delayed and, thus, the deviation from equilibrium is enlarged by increased undercooling. As a result the growth rate of the relative nucleated solid phase increases too (Fig. 5.8a).

Finally, in certain systems the melt viscosity becomes rapidly so large with undercooling that the material transport toward the nuclei by diffusion is almost impossible and the system drives more and more away from equilibrium. In the end, a phase transition of first-order no longer takes place and a second-order transfer of *glass formation* does occur. *Glicksman* (2011) combined both opposite processes in the following viscosity-dependent growth velocity

$$v = \frac{\Delta h}{6\pi\eta r}\frac{\xi_{hkl}}{\Lambda_{hkl}}\frac{\Delta T}{T_e}\exp\left(-\frac{\Delta E_\eta}{k(\Delta T)}\right) \tag{5.20}$$

where Δh is the enthalpy of fusion, η the dynamic viscosity, r the gyration radius of the diffusing atoms/molecules, ξ_{hkl} a crystallographic factor varying slightly with the interface growth direction, Λ_{hkl} a proportionality factor for the width of the solid–liquid transition zone, ΔE_η the activation energy of viscous flow, and T_e the equilibrium temperature to be equated with T_m the melting point of the given substance. The growth rate dependence for attachment-limited crystal growth from a high-temperature viscous silicate melt according eq. (5.20) is constructed in Fig. 5.8 *b*. As can be seen, initially the interface speed (averaged from some glass data) is rising with melt supercooling, like in *a*, but followed by a slowing down at $\Delta T > 200$ K as the viscosity increases at an exponential rate. Thus, when strong viscosity effects occur, a *melt–glass transition* without nucleation and growth of solid phase is approaching.

In this context it is important to note that the classical driving force of the melt–solid transition in eq. (5.2) $\Delta\mu_{L\to S} = \Delta h(\Delta T/T_m) = \Delta s \Delta T$ and its insertion into the work of nucleation in eq. (5.14) is an approximation strictly valid for small and middle supercoolings only. There are material systems for which at a critical large undercooling temperature the viscosity has risen up to such a high value that the diffusion of the atoms or molecules to a possible nucleation position is quasi-stopped. With furthermore decreasing temperature the metastable state of the disordered melt structure is solidified without nucleation of the crystalline phase, which means that the driving force is quasi-annulled. The end result of such transition is the *glass formation* characterized by "freezing of the undercooled melt state". After *Kauzmann* (1948) at this kind of transition the difference between entropy of the liquid and of the solid $\Delta s = s_l - s_s$ becomes equal to zero.

5.2.1.2 Nonclassical concept

As was shown in Section 5.2.1.1 according to the classical nucleation theory (CNT) the crystal nuclei are born randomly and their growth rate is determined by the driving force of crystallization $\Delta\mu$ at a fixed interface energy γ (eq. (5.18)). From the specification of the pre-factor of the nucleation rate in eqs. (5.16) and (5.19), it became apparent that the diffusion constant D and its dependence on the viscosity η of the fluid phase play a decisive role. However, the classical theory ignores an ordered crystal-like pre-orientation even in markedly supercooled fluids, but also, if there are strong bonding relationships between the atoms or molecules. Furthermore, the basic assumptions of CNT are questionable as they do not consider the microscopic nature of the transition. One of the recognized sources of uncertainty is given by the assumption of the capillary approximation considering the surface energy γ as a fixed value independent of the nucleus size that refers to a flat interface. Although valid for critical clusters of large size, it loses consistency for small clusters.

Especially during the last few decades several studies have reported nonclassical nucleation mechanisms characterized by the initial formation of pre-ordered regions in the liquid that act as precursors of the crystallization and related polymorphic structures. The clusters of the pre-structured liquid are regions of either increased bond-orientational order or density that promotes the emergence of crystallites within the center of the clusters by reducing the interfacial free energy. Thus, the observation, e.g., by HRTEM and X-ray scattering (see lecture part II), and numeric MD modeling of preordered regions in supercooled melts has raised great interest in understanding the connection between structural and dynamical heterogeneity of the liquids and crystallization mechanisms in a diverse range of metallic, semiconducting, dielectric, and organic systems.

The nonclassical concept analyzes the nucleation mechanism via two steps: the initial formation of preordered regions in the supercooled liquid and a subsequent formation of the crystalline bulk phase within these regions. The increase in the bond-orientational order within the preordered liquid region creates a precursor that reduces the interfacial free energy by providing a *diffuse interface* between the liquid and crystalline core of the growing nucleus, thus facilitating the formation of the bulk phase. A key component in the analysis of the simulation turns out the identification of the local structure around each atom typically defined on the basis of *order parameters* (OPs) – functions of the Cartesian coordinates that relate a numerical value to a spatial configuration of an ensemble of atoms or molecules. An OP is typically zero for a disordered phase while it assumes characteristic nonzero values for specific spatial ordered arrangements. Recently, *Mahata et al.* (2022) analyzed by using the *common neighbor analysis (CNA)* via the *Steinhardt* OP[1]) the influence of liquid preordering on homogeneous nucleation of pure metals and compared the results with CNT. The obtained critical nucleus sizes of Al and Mg within own high-supercooled melts was markedly reduced up to ~50% compared to the classical pathways.

In a review paper, *Vekilov* (2010) shows the importance of consideration of the preordering effect at the growth from solutions. The most significant finding is the *two-step mechanism of nucleation*, according to which the crystalline nucleus appears inside preexisting metastable clusters of size several hundred nanometers, which consist of dense liquid and are suspended in the solution. While initially proposed for protein crystal nucleation, the applicability of this mechanism has been demonstrated for small molecule organic and inorganic materials, colloids, and biominerals too.

Again, for a more special insight into the theory, please refer to the extensive literature. Suffice it to point out here that we can no longer neglect the effect of a possible pre-orientation in the liquid. In the lecture part II we will come back to this phe-

---

1 Steinhardt OP: a set of parameters based on spherical harmonics to explore the local atomic environment. Such parameters are used to identify structures of solid and liquid atoms (see *Steinhardt et al., 1983*).

nomenon at the discussion of the kinetics of growing liquid–solid interfaces showing often not a sharp boundary but a diffuse transition region consisting of few pre-oriented atomic rows in the melt with features of the crystal structure to be grown.

5.2.2 Heterogeneous nucleation

For the growth of single crystalline bulk and thin-film materials the homogeneous nucleation does not play an important role. On the contrary, it should be even avoided in order to obtain monocrystalline structures without disoriented multinucleated grains. Thus, for bulk crystal growth the use of an *artificial seed* is favored to start the crystallization locally and reduce the driving force of crystallization by the presence of a supporting fluid–solid interface. At epitaxial processes a *substrate* acts as quasi-seeding surface area fitting the structure of the depositing layer as much as possible. Although in these cases nucleation processes are furthermore of decisional role they are now of *heterogeneous character*, which reduces the barrier of activation energy and enhances the controllability. However, besides a controllable nucleation step, it may also occur that during crystal growth and epitaxy uncontrollable heterogeneous contacts of the starting phase with the walls of containers, ampoules, and crucibles take place, which leads to unwanted heterogeneous nucleation. Let us quantify the energy difference between homogeneous and heterogeneous nucleation.

5.2.2.1 Basic considerations

The classical theory for heterogeneous nucleation again assumes the simple droplet model, but now in contact with a foreign solid area (substrate). Such interaction with a solid underlay influences the shape of the droplet. Dependending on the contact angle θ, also known as *wetting angle*, a cap-shaped *spherical segment* is formed (see Fig. 5.9). This is due to the reduced work of nucleation introduced in eq. (5.14). Similar to the situation at a crystal–liquid–vapor triple contour at melt growth that was discussed in Section 4.4.1, also here interface energies are in thermodynamic equilibrium, namely between: (i) solid substrate and vapor (SV), (ii) substrate and liquid droplet (*SL*), and (iii) liquid droplet surface and vapor (*LV*). They are linked to one another via the wetting angle θ as given by Young's equation

$$\cos\theta = \frac{\gamma_{SV} - \gamma_{SL}}{\gamma_{LV}} \tag{5.21}$$

with the interface energies between solid substrate and vapor γ_{SV}, solid and liquid γ_{SL}, and liquid and vapor γ_{LV}, respectively. Depending on the degree of wetting the value of θ lies in the range of 0° (total wetting) to 180° (quasi-non-wetting). Again, the free enthalpy of such spherical segment with radius r_{seg} is composed of the volume and interface-related enthalpy parts. However, now the volume of the spherical segment is

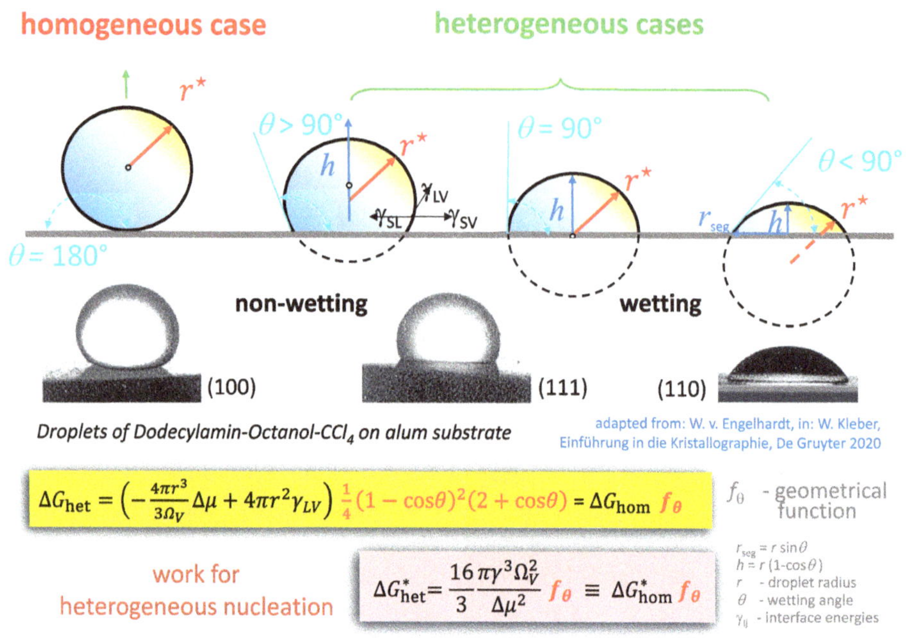

Fig. 5.9: Sketched and experimentally proven correlation between non-wetting and wetting of a sphere and spherical droplet segments, respectively, showing the difference between homogeneous and heterogeneous nucleation due to varying wetting angles on a substrate.

$V_{seg} = \frac{4}{3}\pi r^3 \frac{1}{4}(1-\cos\theta)^2(2+\cos\theta)$ and the droplet surface consists of: (i) the free interface with vapor $F_{LV} = 2\pi\, r\, h$ and (ii) the contact interface $F_{SL} = \pi r_{seg}^2$ with the substrate. After inserting the spherical segment height $h = r(1-\cos\theta)$ and the radius of the contact area of the spherical segment $r_{seg} = r\,\sin\theta$, with r the radius of the segment curvature (see Fig. 5.9), the Gibbs free energy change at heterogeneous nucleation is

$$\Delta G_{het} = -\left[\frac{4\pi r^3}{3\Omega_V}\frac{1}{4}(1-\cos\theta)^2(2+\cos\theta)\right]\Delta\mu + 2\pi r^2\,(1-\cos\theta)\gamma_{lv} + \pi r^2 \sin^2\theta(\gamma_{SL} - \gamma_{SV}) \tag{5.22}$$

Note that whereas the interface area between the droplet and vapor is multiplied by the new created (!) interface energy γ_{LV} (second term on the right side) the contact area between droplet and substrate is not required to form a totally new interface energy γ_{SV} but uses partially the already existing boundary energy between vapor and substrate $\gamma_{SV} > 0$. Thus, in the third term the nucleated contact area between droplet and substrate needs to be multiplied by the reduced value $(\gamma_{SL} - \gamma_{SV})$.

After some trigonometric transformations of eq. (5.22) we get the expression

$$\Delta G_{\text{het}} = \left(-\frac{4\pi r^3}{3\Omega_V}\Delta\mu + 4\pi r^2\gamma_{LV}\right)\frac{1}{4}(1-\cos\theta)^2(2+\cos\theta) = \left(-\frac{4\pi r^3}{3\Omega_V}\Delta\mu + 4\pi r^2\gamma_{LV}\right)f_\theta \quad (5.23)$$

with the geometrical function f_θ, mostly named *wetting function*, in which the heterogeneous case differs from the homogeneous one [compare with eq. (5.11)].

Of course, the wetting behavior on a crystalline substrate is crystallographically determined due to the anisotropy of the interface energy. Depending on the orientation of the substrate surface the wetting angle of a droplet of a given substance can differ markedly. Examples are liquid droplets of dodecylamin: octanol: CCl_4 on the (100), (110), and (111) surfaces of an alaun substrate the images of which are shown in Fig. 5.9. Whereas the (100) plane wets almost totally the (110) surface behaves like contact-free.

The critical nucleus radius r^* is in principle obtained in the same way as for homogeneous nucleation by deriving and zeroing the eq. (5.23)

$$\frac{\partial \Delta G_{\text{het}}}{\partial r} = \left(-\frac{4\pi r^2}{\Omega_V}\Delta\mu + 8\pi r\gamma_{LV}\right)f_\theta = 0 \quad (5.24)$$

$$r^*_{\text{het}} = \frac{2\gamma_{LV}\Omega_V}{\Delta\mu} \equiv r^*_{\text{hom}}\,|_{\gamma_{LV}=\gamma} \quad (5.25)$$

where the radius of heterogeneous nucleation represents the radius of curvature of the spherical droplet segment (red r^* in Fig. 5.9). Therefore, the critical nucleation radius is basically the same for heterogeneous nucleation as for homogeneous nucleation, only the activation energy differs (under otherwise identical conditions). Insertion of the critical radius r^*_{het} in eq. (5.23) yields the critical activation energy (work of nucleation) of heterogeneous nucleation

$$\Delta G^*_{\text{het}} = \frac{16}{3}\frac{\pi\gamma^3_{LV}\Omega^2_V}{\Delta\mu^2} f_\theta \equiv \Delta G^*_{\text{hom}}\, f_\theta \quad (5.26)$$

with $f_\theta = \frac{1}{4}(1-\cos\theta)^2(2+\cos\theta) \le 1$, which means that the heterogeneous nucleation requires less activation energy than the homogeneous nucleation.

Figure 5.10a shows the graphic dependence of the function f_θ on the wetting angle θ. If the wetting angle approaches 180° so that the function $f_\theta \approx 1$ the spherical cap becomes a complete sphere that no longer wets the substrate or container wall. Then the case of quasi-homogeneous nucleation is obtained. Conversely, at strong wetting with very small wetting angle, f_θ decreases toward zero and the activation energy required for nucleation in eq. (5.23) is greatly reduced. In between when the wetting angle is around 90° the nucleus is of hemispheric shape and the function $f_\theta \approx 0.5$. In this case, the nucleus requires only half the activation energy. As is sketched in Fig. 5.10a it has only half as much volume as a spherical nucleus during homogeneous nucleation. Usually, the situations with $\theta < 90°$ are classified as good wettable, as in the cases of silicon droplets on Si_3N_4 (~20°), graphite (30°) and only just on fused silica (87°). On the other hand values of $\theta > 90°$ are referred to non-wetting like molten GaAs on pBN (155°) or CdTe on graphitized fused silica (100°). The special case of so-called spreading at $\theta = 0°$ with $f_\theta = 0$ occurs at ideal epitaxy designated as *homoepitaxy* (not to be confused with homogeneous nucleation being exactly the opposite process). Figure 5.10b shows the free energy change ΔG (nucleation work) and the related critical values ΔG^* of homogeneous and heterogeneous nucleation at various wetting angles. As an example, a droplet of high-purity liquid iron in contact with hypothetically different wetting surfaces is taken. The following Fe parameters of its molten state have been applied: $T_m = 1803$ K, $\Delta h = 13.8 \times 10^3$ J/mol, $\gamma_{LV} = 1.87 \times 10^{-4}$ J/cm^2, $\Omega_V = m\ (N_A \rho)^{-1} = 1.15 \times 10^{-23}$ cm^3 with m and ρ the molar mass and density, respectively. Equation (5.2) deviated by the Avogadro number was taken as driving force, i.e., $\Delta\mu = \Delta h\ N_A{}^{-1}\ \Delta T\ T_m{}^{-1} = 0.38 \times 10^{-20}$ J. First the homogeneous case was estimated by using eqs. (5.25) and (5.26) and setting $\theta = 180°$ into the f_θ function. At maximal observed undercooling of liquid iron $\Delta T_{max} = 295$ K (see *Table 5.1*) critical nucleation radius r^* and related nucleation work ΔG^* yield about 11.4 nm and 10.2×10^{-16} J, respectively. With decreasing wetting angle, the critical nucleation work decreases drastically (Fig. 5.10b). In reality, the solid nucleus formation in liquid iron plays an important role in casting processes where the grain texture of the ingot is determined by the heterogeneous nucleation on a given mold wall (note, then the applied L-V surface tension used so far should be replaced by the S-L interface energy, which is $\gamma_{SL} \approx 0.2 \times 10^{-4}$ J/cm^2).

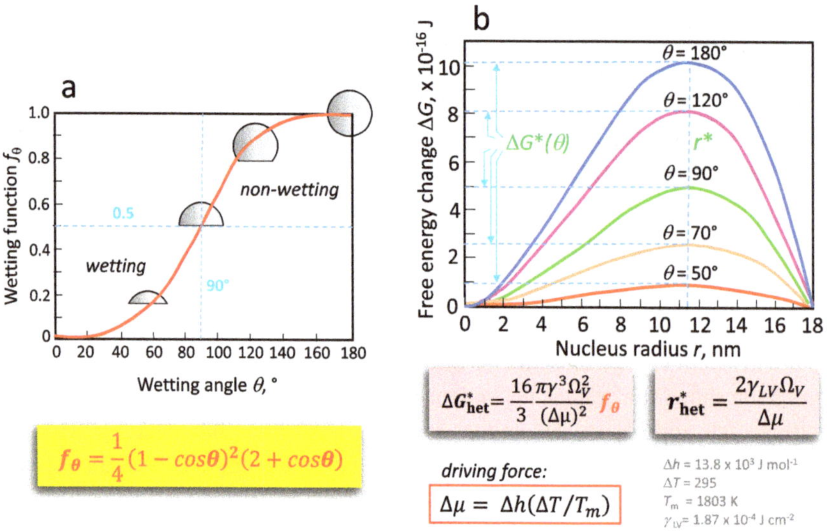

Fig. 5.10: Calculated wetting function f_θ (θ) (a) and free energy change ΔG versus nucleus radius r varying the wetting angles θ from 0° to 180° of a liquid iron droplet (b).

5.2.2.2 Application in epitaxial processes

In practice, in bulk single crystal growth and thin-film epitaxy, nucleation is initiated on a monocrystalline seed or substrate. When a seed or substrate is used whose crystal structure is identical to the fluid material to be crystallized or deposited, in other words when the case of total wetting takes place, a two-dimensional (2D) homoepitaxial nucleation is favored. In first approximation the height of such a nucleus is equal to the lattice parameter of the given substance. After one 2D nucleus is formed, it would be preferable if the whole interface plane is then rapidly completed by lateral growth. However, as we will show below, in reality, the statistical character of the nucleation process generates simultaneously many 2D nuclei coalescing during their lateral expansion. But first, for simplicity, a single solid *disk-shaped nucleus* is formed on a homoepitaxial substrate in a supersaturated vapor phase (Fig. 5.11, above left). Again, the nucleus activation energy is composed of the negative volume part and positive surface term as

$$\Delta G_{\text{disc}} = -\frac{\pi r^2 a}{\Omega_V} \Delta\mu + 2\pi r a \gamma \qquad (5.27)$$

critical nucleus radius:

$$r^*_{\text{disc}} = \frac{\gamma \Omega_V}{\Delta\mu}$$

equilibrium ratio:

$$\frac{h}{l} = \frac{\gamma_{nv} + \gamma_{ns} - \gamma_{vs}}{2\gamma_{nv}}$$

γ_{nv} – interface energy between nucleus and vapour
γ_{ns} – interface energy between nucleus and substrate
γ_{vs} – interface energy between vapour and substrate

2D disc-shaped nucleus on homoepitaxial substrate

3D rectangular nucleus on heteroepitaxial substrate

1) $\gamma_{vs} < \gamma_{nv} + \gamma_{ns}$
$h > 0$
- non-wetting substrate
- certain nucleus height is required for growth
→ **3D nucleation mode**

Volmer-Weber

2) $\gamma_{vs} > \gamma_{nv} + \gamma_{ns}$
quasi $h < 0$
- wetting substrate
- lateral nucleus spreading becomes dominant
→ **2D nucleation mode**

reduced attraction

Frank-van der Merwe

3) $\gamma_{vs} = \gamma_{nv} + \gamma_{ns}$
$h = 0$
- mono-atomic layer-by-layer spreading
- after $h_\Sigma > h^*$ island growth can be occurred

monolayer

2D
↓
3D

Stranski-Krastanov

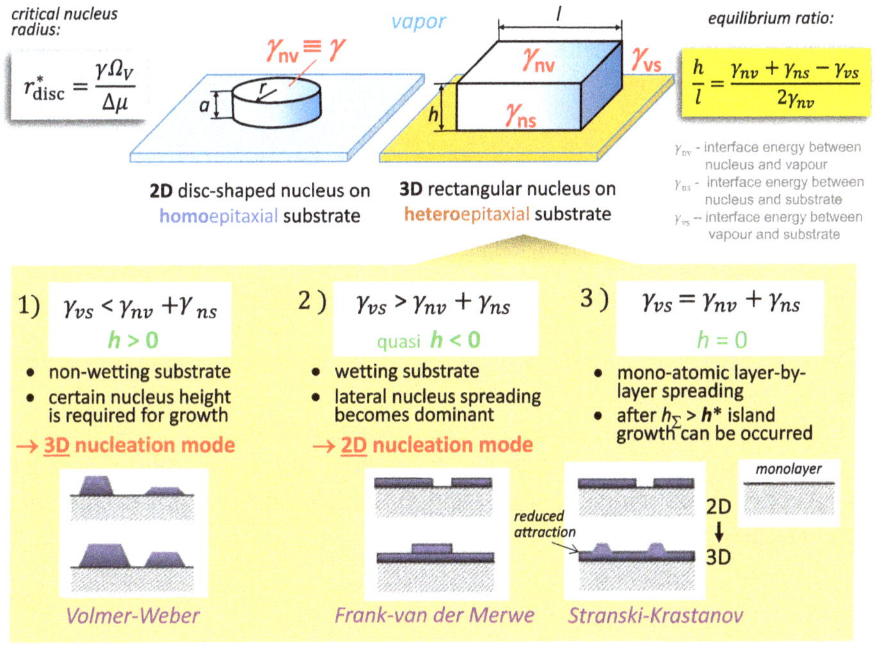

Fig. 5.11: 2D and 3D nucleation modes on substrates from vapor determined by the interplay of the three interface energies.

with r the disk radius, a the height of the disk to be equated at first approximation with the dimension of the growth unit (atom, molecule) of the given nucleus material, and γ the surface energy, which is now, however, exclusively acting between the solid seed or substrate and the surrounding vapor. Thus, no interface related term of the nucleus base has to be considered in eq. (5.27) due to the absolute structural fitting between nucleus and seed or substrate. After $\partial \Delta g_{\mathrm{disk}}/\partial r = 0$ the critical nucleus radius is

$$r^*_{\mathrm{disk}} = \frac{\gamma \Omega_V}{\Delta \mu} \tag{5.28}$$

and the critical nucleation energy for a 2D disk-shaped *homoepitaxial nucleus* is then

$$\Delta G^*_{\mathrm{disk}} = \frac{\pi a \gamma^2 \Omega_V}{\Delta \mu} \tag{5.29}$$

Replacing the homogeneous droplet and disk models by heterogeneous *polyhedral-shaped 3D nuclei* the situation becomes more realistic regarding the formation of a single crystalline solid phase at epitaxial processes from vapor or melt–solutions. Figure 5.11 above right shows the model of such rectangular 3D nucleus of height h with a quadratic main surface of edge length l. By all means, such a type of nucleus is not unusual, like KCl on NaCl substrate, for example. The formation of such nucleus (n) on a substrate (s) in contact with a vapor phase (v) changes the molar thermodynamic potential of the system as

$$\Delta G_{\mathrm{3D}} = -\frac{l^2 h}{\Omega_V} \Delta \mu + l^2(\gamma_{nv} + \gamma_{ns} - \gamma_{vs}) + 4lh\gamma_{nv} \tag{5.30}$$

with the volume part $-\Delta G_V = (l^2\, h/\Omega_V)\Delta\mu$ and the two interface parts. The first (middle term) represents the change of the free interface energy between substrate and vapor phase when a crystalline cuboid with surface energy γ_{nv} appears on the substrate and creates the interface energy of the contact area between nucleus (n) and substrate γ_{ns} (adhesion energy). The second surface term stands for the formation energy of the side walls.

The work of nucleation of such a cuboid depends on both the absolute dimensions h and l and their ratio h/L (habitus). For a given volume $l^2h = \mathrm{const}$ the work of nucleation proves to be minimal when the sum of the two surface parts in eq. (5.30) is minimum too as

$$\frac{\partial}{\partial L/h}\left[l^2(\gamma_{nv} + \gamma_{ns} - \gamma_{vs}) + 4lh\gamma_{nv}\right] = 0 \tag{5.31}$$

giving the equilibrium shape [relation of *Dupré* (1869)]

$$\frac{h}{l} = \frac{(\gamma_{nv} + \gamma_{ns} - \gamma_{vs})}{2\gamma_{nv}} \tag{5.32}$$

Using eq. (5.32) for replacing in eq. (5.30) h by l or l by h and zeroing the first derivatives $\partial \Delta g_{\mathrm{3D}}/\partial h$ or $\partial \Delta g_{\mathrm{3D}}/\partial l$ becomes the critical edge length $l^* = 4\Omega_V \gamma_{nv}/\Delta\mu$ or height

$h^* = 2\Omega_V \gamma_{nv}/\Delta\mu$, respectively. Then the critical nucleation energy for a heterogeneous rectangular 3D nucleus is

$$\Delta G^*_{3D} = 16 \frac{\gamma_{nv}^2 \Omega_V}{\Delta\mu^2} (\gamma_{nv} + \gamma_{ns} - \gamma_{vs}) \tag{5.33}$$

Based on the actual constellation of the three acting interface energies γ_{ij} the following three striking cases can occur:

i) $\gamma_{vs} < \gamma_{nv} + \gamma_{ns}$ so that after eq. (5.32) $h > 0$: there is only a small (or missing) interaction with the substrate (comparable with non-wetting situation) and a certain nucleus height is required for the growth. Such spatial shaping normally to the substrate surface is defined as *3D nucleation process* or rather as *Volmer–Weber mechanism* (Fig. 5.11–1).

ii) $\gamma_{vs} > \gamma_{nv} + \gamma_{ns}$ so that after eq. (5.32) $h < 0$: there is a significant interaction with a highly attractive substrate (comparable with wetting situation) and the lateral nucleus spreading becomes the determining factor. Such occurrence is referred to as *2D nucleation process* or rather as *Frank–van der Merwe mechanism* (Figs. 5.11–5.12).

iii) $\gamma_{vs} = \gamma_{nv} + \gamma_{ns}$ so that after eq. (5.32) $h = 0$: this is a particular case of *barrier-free layer-by-layer spreading* over the substrate surface. However, beyond a critical layer thickness $h_\Sigma > h^*$ is reached, which depends on strain and chemical potential deposited film, growth continues through nucleation mechanism. Such *layer-plus-island growth* is named *Stranski–Krastanov mode* (Fig. 5.3).

In case of nucleation from a liquid or solution the related interface energies γ_{nf} and γ_{sf} (with index f for a given fluid phase) have been used. More details from a molecular-kinetic point of view will be given in the lecture part II. Here the discussion in Section 4.3 whereupon a nucleated crystallite possesses a certain polyhedral quasi-equilibrium shape determined by the given crystallographic structure should be remembered. As shown in Spec box 5.2, the formation energy of a homogeneous crystalline nucleus must consider the anisotropy of the crystallographic planes A_{hkl} and related surface tensions γ_{hkl} as

$$\Delta G = -n\Delta\mu + \sum \gamma_{hkl} A_{hkl} \tag{5.34}$$

with $n = V_{cr}/\Omega$ the number of atoms or molecules in the nucleus. Of course, at a given supersaturation or supercooling also both the habitus and size of a heterogeneous nucleus have to remain the same Wulff's central distances h_{hkl} proportionally to the corresponding surface energies γ_{hkl}. Thus, eqs. (5.30)–(5.33) have been modified accordingly (the somewhat awkward procedure is here not shown but given in detail in the related literature).

It is noteworthy that at the crystal growth from melt mainly a quasi-nucleationless crystallization due to the very small supercooling (i.e., driving force $\Delta\mu$) takes place. This also happens at the propagating crystallization front due to the characteristic atomic roughness of a melt–solid interface (see lecture part II). One exception is the presence of facets (i.e., atomically smooth faces) requiring more driving force and, thus, a possible nucleation-assisted mode (see Section 4.4.4).

5.2.2.3 Nonequilibrium nucleus distribution

First *Volmer* assumed the probability of finding the equilibrium critical cluster number N^* consisting of critical number n^* of atoms or molecules [compare with eq. (5.18)] among the existing clusters as

$$N_i^* = N_i \ exp\left(\frac{-\Delta G^*}{kT}\right) \tag{5.35}$$

with N_i the number of clusters consisting of $i = 1, 2, 3, \ldots, n$ atoms or molecules referred to as monomers, dimers, trimers, etc. Such a situation for the heterogeneous case is sketched in Fig. 5.12a. There are disk-shaped clusters on a substrate within a supersaturated fluid (e.g., vapor) having still an undercritical radius $r < r^*$ and possessing, thus, a high decay probability. At the same time a critical nucleus with $r = r^*$ (in red) and an overcritical nucleus with $r > r^*$ are added. *Volmer's* related equilibrium distribution $N_i(i)$ is shown by curve 1 in Fig. 5.12a. The number of critical nuclei N_i^* is the lowest due to both decay and growth as equitable probabilities. At $i > i^*$ only growth without reverse reaction is expected. As a result the number of overcritical (growing) nuclei should be assumed as increasing as drafted by the dashed segment in curve 1. However, this assumption proved to be incorrect and must be modified by considering kinetic processes that deviate the situation toward nonequilibrium distribution (demonstrating the close connection between thermodynamics and kinetics).

In 1935 *Becker and Döring* extended the treatment of the nucleation process by considering the kinetic actions and related time behavior of the cluster distribution. Accordingly, the clusters can grow and decay by attachment or detachment of single atoms or molecules. As a result, the molar activation energy Δg for cluster formation is a time-varying term and reduces by the amount $\Delta g_n - \Delta g_{n-1}$ when an atom is removed, for example. Thus, an expression for the *rate of nucleation* considering growth and decay had to be formulated. With the assumption that a cluster quantity N_i is formed only by adding an atom to a cluster N_{n-1} or by leaving a cluster N_{n+1} and is destroyed if it either gains or loses an atom, the time dependence of the cluster concentration is

$$\frac{dN_i}{dt} = J_N - J_{N+1} = \left(\upsilon_{n-1}^+ N_{n-1} + \upsilon_{n+1}^- N_{n+1}\right) - \left(\upsilon_n^+ c_n + \upsilon_n^- N_n\right) \tag{5.36}$$

with $J_N = \upsilon_{n-1}^+ N_{n-1} - \upsilon_n\text{-}N_n$ the net flux of clusters through the sizes n and $n + 1$, and, υ_n^+ and υ_n^- the probabilities for attachment and detachment of single atoms or molecules, respectively.

a

γ

r^*

a

$r > r^*$

$r < r^*$

$r < r^*$

Cluster number N_i

1 *Volmer's* equilibrium function

$N_i^* = N_i \, exp\left(\frac{-\Delta G^*}{kT}\right)$

2

N_i^* eq

N_i^* noneq

Becker–Döring's non-equilibrium function

$N_i^* = Z \, N_i \, exp\left(\frac{-\Delta G^*}{kT}\right)$

$n_1 \; n_2 \; ...$ $\quad n_i^*$ Species n_i

b

nucleation rate:

$$J = Zv^{+*}N_i^* = Zv^{+*}N_i \, exp\left(\frac{-\Delta G^*}{kT}\right)$$

Z - Zeldowich factor
v^{+*} - attachment probability of an atom on a cluster
N_i^* - number of critical nuclei consisting of n_i^*
N_i - cluster concentration
ΔG^* - critical nucleation energy $\sim \Delta\mu^{-1}$

J

heterogeneous cases

homogeneous case

$\Delta\mu^*$

$\Delta\mu$

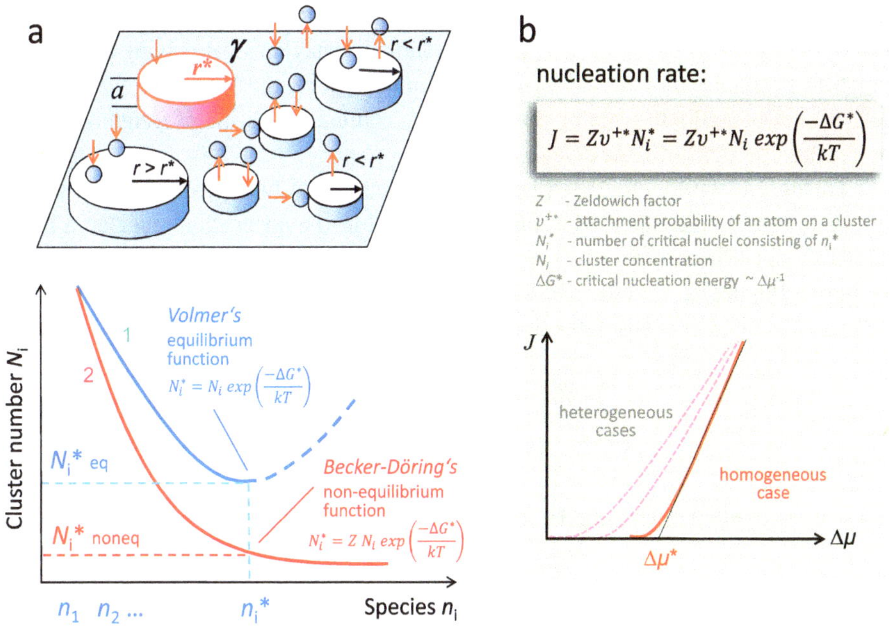

Fig. 5.12: Becker–Döring's nonequilibrium compared with Volmer's equilibrium nucleus number (a) and nucleation rate versus chemical potential (b).

Without entering the kinetic details the multiplication of the quantity of nonequilibrium critical nuclei $N_i^* = Z \, N_i$ with the constant attachment probability to the nucleus v_n^{+*} = const yields the steady-state nucleation rate

$$J = Zv^{+*}N_i^* = Zv^{+*}N_i \, \exp\left(\frac{-\Delta G^*}{kT}\right) \tag{5.37}$$

with the *Zeldovich factor* $Z = (1/n^*)(\Delta g^*/3\pi kT)^{1/2} \geq 10^{-2}$ where n^* is the number of atoms/molecules in the cluster and Δg^* is the formation energy of the cluster. Thus, Z gives the probability that a nucleus at the top of the barrier will go on to form the new phase, rather than dissolve. Adding $N_i^* = Z \, N_i$ in Fig. 5.12a as curve 2, a quantitatively lowered course compared to curve 1 is obtained. Then, the nonequilibrium of the process is characterized by both steady primary growth and possible decay continuing even after the critical nucleus creation. A quasi-mutual delivery and removal of species from each other takes place whereas the larger nuclei are favored compared to the smaller ones due to their smaller supersaturation known from the Gibbs–Thomson effect [see eq. (4.10)]. They are growing at the expense of smaller ones. Finally, only a few "winners" of largest size are remaining through coalescence (see also the following chapter).

In Fig. 5.12b the nucleation rate as function on the driving force (supersaturation) according to eq. (5.19) is sketched in order to illustrate its exponential character. For that purpose the indirect proportionality between the activation energy and driving force $\Delta G \sim \Delta\mu^{-1}$ was considered. A certain potential delay (critical driving force $\Delta\mu^*$) must be overcome in order to form critical nuclei. Sure, the better the wetting between nucleus and substrate or the higher the content of assisting foreign atoms or molecules, the shorter is the delay.

Further, as will be discussed in the lecture part II, at heterogeneous nucleation, the *desorption* and *surface diffusion* of atoms within the adsorption layer on a used substrate must be considered. Then the exponential factor in eq. (5.37) is modified by including the activation energies for desorption E_d and diffusion $-E_s$ so that the following completed expression for the nucleation rate is obtained

$$J = Zv^{+*} N_i \exp\left(\frac{E_d - E_s - \Delta G^*}{kT}\right) \tag{5.38}$$

5.2.3 Uncontrolled nucleation in crystal growth containers

Today, the unidirectional crystallization within a container or ampoule starting from an artificial seed at the bottom proves to be one of the most effective production methods of single crystals. The corresponding VB and VGF techniques are successfully applied to grow many semiconductor compounds (e.g., GaAs, InP, and CdTe), oxides and organic crystals of high quality. Due to the high diameter constancy and low thermal inhomogeneities (small linear temperature gradients) a relative low defect density can be obtained. However, there arises a not yet completely solved problem of uncontrolled nucleation at the inner container wall creating disoriented grains, dislocations, and twins, which penetrate into the growing crystal. The origin proves to be the contradiction between the desired non-wetting of the container wall and usually concave isotherm courses. This will be explained in detail in the following.

First, an essential precondition for high-quality crystal growth in containers is the non-wetting behavior between the melt and inner container wall to reduce the uncontrolled heterogeneous nucleation probability [see eq. (5.23)]. For this purpose either a suitable container material is found or the wall is covered by a non-wetting thin-film coating (e.g., by pyrolytic graphitization). Sure, the coating film must be strongly cohesive. On the other hand, there is a characteristic difference of thermal conductivity in the melt and solid $\lambda_L \neq \lambda_S$. Especially in semiconductors λ_L is about two times higher than λ_S in crystalline phase. In Fig. 5.13 the thermal conductivities of important semiconductors are tabulated. In consequence, the shape of the propagating solid–liquid interface (IF) becomes concave. This is due to the heat "holdup" during its transport from the hot region of the melt toward the IF whereupon the lower thermal conductivity within the growing crystal impedes the flux of the arriving heat quantity. Consequently, the path of the heat flux is changed toward radial direction, creating by this means a concave course of the melting point isotherm (see Fig. 5.13, left). Thus, the IF will be concave-curved too. However, from mesoscopic point of view this is only valid until the wall proximity because in this region the IF will be convex recurved due to the non-wetting situation at the container wall (magnified situation in the middle of Fig. 5.13). Some wetting angles on typical container walls are given in the table below right. As a result a ring area of enhanced undercooling (driving force) between the convex IF rim and concave melting point isotherm is evolved. This may lead to unwanted nucleation in this region with the abovementioned consequences. Such effect is being further exacerbated by the presence of facets at the interface periphery, an aspect already discussed in Section 4.4.3. It was shown that the facets represent the planes of lowest free interface energy and, thus, of highest undercooling. In the course of interaction between melt, facet, and container wall, the disoriented nucleation and even twinning probability are enhanced.

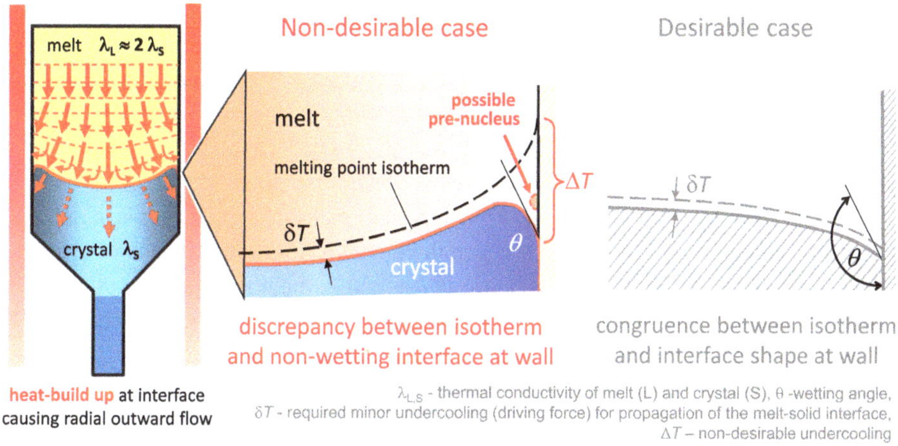

***Thermal conductivity** in melt (λ_L) and solid (λ_S)*

Material	Si	Ge	GaAs	InP	CdTe	ZnTe
λ_L W/cmK	0.56	0.36	0.18	0.23	0.02	0.04
λ_S W/cmK	0.19	0.16	0.07	0.09	0.01	0.02

adapted from:
M. Jurisch et al. in: HB of Crystal Growth IIA (Elsevier) 331

***Wetting angle** θ on typical crucible walls*

Material	Si	Ge	GaAs	GaAs	CdTe
Container	graphite	graphite	SiO_2	pBN	pBN
θ °	15	135	120	170	155

adapted from:
Th. Duffar in: Handbool of Crystal Growth IIB (Elsevier) 757

Fig. 5.13: Wrong nucleation within the undercooled region at the inner container wall due to concave melting isotherm and non-wetting of the melt at VB/VGF growth to be avoided by a convex isotherm (*with permission of Elsevier (Tables)*).

The best means to counter is to ensure a near-flat or slightly convex interface shape (as shown in Fig. 5.13, right) that proves to be a serious task for the hot zone design.

5.2.4 Precipitation in cooling crystals

The formation of precipitates of foreign phases in an as-grown crystal during its cooling down to the room temperature is a *solid–solid nucleation* process taking place at enhanced concentrations of an added component due to its characteristic retrograde solubility. For instance, in Fig. 3.28b the retrograde solubility curves of impurities/dopants in silicon crystals are shown. It is a well-known serious problem of oxide precipitation in silicon Czochralski crystals because of the relative high oxygen concentration within the melt released from the fused silica container and incorporated into the growing crystal. While at temperatures near the melting point the oxygen solubility is still relative high (~10^{18} cm^{-3}) during cooling to 600 °C it falls below 10^{16} cm^{-3}. As a consequence microscopically SiO_x precipitate modification (crystoballit, quartz) can be formed. That means in such a case the nucleation is combined with a chemical reaction between the matrix (Si) and the foreign substance (O). Further, oxygen precipitates in CZ silicon crystals are growing on clusters of excess vacancies quasi-heterogeneously. Also other impurities, in particular fast diffusing metals, like Cu or Au, may be heterogeneously precipitated at intrinsic point defect aggregates like dislocation loops (swirls) but apart from a chemically reactive process with silicon. It has been observed that already an Au content of about 10^{13} cm^{-3} is sufficient to result in such a preferential Au precipitation.

Precipitation plays an evident role in binary and multicompound crystals of nonstoichiometric composition. It is connected with the excess concentration of one of the components, which is at high temperatures still solved within the matrix as interstitial or vacancy atoms. As was shown in the Figs. 3.13, 3.14, and 3.28a also the intrinsic point defect solubility is decreasing with reducing temperature, clearly visible in the retrograde solidus curves of the homogeneity regions (see Sections 3.2.4.2.1–3.2.4.2.2). Again, during the as-grown crystal is cooling down the solidus is crossed and nucleation of second-phase particles or microvoids takes place via interstitial or vacancy conglomeration, respectively. Thus, one has to differ between *homogeneous* or *heterogeneous precipitation* within the compound matrix or at the contact with a supporting another defect aggregate (e.g., dislocations and grain boundaries), respectively. Usually, the heterogeneous case requires much lower nucleation energy and is therefore the favored type. Whereas homogeneously nucleated precipitates show a very small size of some 10–100 nm and their analysis requires high-resolution electron microscopy, heterogeneously formed precipitates decorating dislocations are well detectable by IR transmission microscopy. Depending on the degree of deviation from stoichiometry average precipitate densities in the region of 10^8–10^{12} cm^{-3} have been found in arsenic-rich GaAs and Te-rich CdTe, for example. Principally, in all materials such second-phase particles impair not only the wafer polishing and epitaxy processes but also the transmission quality by light scattering of optical devices made from such wafers. Measures of their minimization are near-stoichiometric growth conditions (Section *3.2.4.2.3*) and wafer annealing.

Generally, the nucleation of precipitates can be treated analogously to the nucleation of a droplet in a vapor or a nucleus within the melt (see Section 5.2.1). However, an additional complication now arises, caused by the small change of the matrix volume, which must be accommodated elastically leading to a strain effect, which has to be considered in the volume part of nucleation free energy $-\Delta G_V$. The appearing elastic energy E_{el} increases with growing precipitate volume V_{pr} via strain energy ε_{el} per volume unit as $E_{el} = \varepsilon_{el} V_{pr}$. For the *homogeneous case*, eq. (5.11) becomes

$$\Delta G = \frac{4\pi r^3}{3}(-\Delta\mu/\Omega_V + \varepsilon_{el}) + 4\pi r^2 \gamma \tag{5.39}$$

where r is the nucleus radius, $\Delta\mu$ is the driving force, Ω_V is the specific volume of the building blocks (atoms and molecules), and γ is the specific solid–solid interface energy between the matrix α and precipitating phase β. Accordingly, the critical precipitate radius of the $\Delta G(r)$ curve is

$$r^* = \frac{2\gamma}{(\Delta\mu/\Omega_V - \varepsilon_{el})} \tag{5.40}$$

That means the ratio between the α–β interface energy and strain energy begins to be a determining factor. In the simplest case, the elastic strain depends on the relative misfit between the lattice parameters of both solid phases $\delta = (a_\alpha - a_\beta)/1/2\,(a_\alpha + a_\beta)$ and the (isotropically assumed) elastic modulus of the matrix E_α. When the elastic strain acts in the matrix only it can be written as

$$\varepsilon_{el}^{\alpha} = \frac{E_\alpha}{1-\nu_\alpha}\delta^2 f\left(\frac{b}{a}\right) \tag{5.41}$$

where ν_α is the Poisson's ratio of the matrix and $f(a/b)$ a shape factor for the axis lengths of a simplified rotational ellipsoidal precipitating nucleus. From eqs. (eq. 5.39) and (5.40) it follows that at only slightly differing lattice parameters ($\delta \to 0$) the elastic strain can be neglected and the precipitate has the spherical shape, showing the lowest product between interface energy and area. This case is referred to as *coherent precipitation.* Against it at markedly differing lattice parameters an *incoherent case* will occur where the shape factor comes into action and a compromise between minimization of the elastic strain and interface energy is taking place. From eq. (5.41), it follows that at constant precipitate volume $\varepsilon_{el}^{\alpha}$ decreases in accordance with the order of sphere–rod–disk. This ranking often corresponds with the growth period of precipitates. Whereas at both small volumes and δ values spherical or needle-shaped forms are preferred, with increasing volume the disk shape prevails due to the $\varepsilon_{el}^{\alpha}$ minimization.

Of course, the detailed process dynamics and kinetics of precipitation proves to be much more complex. First, the crystallographic structures of both solid phases must be considered. It can be assumed that the interfacial energy corresponds to a local minimum when the precipitation reaction realizes a singular interface. Mostly the structural misfit leads to an enormous elastic deformation field around the precipitates generating dislocations. In fact, a marked enrichment of dislocations is reported in many crystals around second-phase particles. Finally, in nonstoichiometric compound crystals the precipitates may be first nucleated in molten state (as droplets) when they consist of an excess component with lower melting point than the matrix. Thus, the precipitate runs through the solidification process during cooling down evoking a delayed structural stress effect. For deeper information on precipitation fundamentals the interested readers are referred to the relevant rich literature and planned lecture part IV on defects.

However, one more important fact must be added. Each precipitation is subjected to the diffusive transport of its building units. In case of a homogeneous distribution of an intrinsic kind of point defects (interstitials or vacancies) within the cooling crystal the sites of their precipitation are distributed uniformly as well. The growth of the just generated nuclei requires a permanent feeding diffusion from adjacencies, i.e., of interstitial atoms or vacancies for precipitates or microvoids, respectively. In the end, the ripening size of both configurations is limited and depends on the relative distance to one another. The growth is completed when the diffusion spheres of each precipitate are exhausted and overlaying. The final size of precipitates with diffusion spheres (expressed as radius of assumed spherical shapes) overlapping on another were fundamentally derived by *Ham* (1958) and specialist by *Wilkes* (1983) for silicon oxide precipitation in silicon crystals. Hence, typical sizes of homogeneously nucleated precipitates are of the order of 10–100 nm, seldom reaching one µm. This is a characteristic feature that makes precipitates differ from inclusions, which are incorporated during crystal growth at the propagating interface and showing dimensions of some µm (as the author found out at growth of CdTe crystals from tellurium-rich melt, for example). Of course, if a heterogeneous precipitation by decoration of presented dislocations occurs, a precipitation-free area around the dislocations is formed. Due to the lower heterogeneous nucleation energy (see Section 5.2.2) the building units are already consumed before homogeneous nuclei are generated. Finally, the precipitation density and size depends on the temperature (diffusivity) and cooling rate too. As is well-known, these relations are summarized in so-called *TTT (temperature–time–transformation) diagrams.*

5.3 Ostwald ripening and grain coarsening

At a free multinucleation process without artificial nucleation, when an intentional nucleus selection is missing or a too high supersaturation is acting, or when too fast cooling and crystallization rates are used, then a multigrain structure is formed. It consists of coalesced crystallites, after stochastic nucleation, which show varying crystallographic orientations. Although such a texture is typical and is of importance in metallurgical processes, such as casting, it has little to do with our lecture topic – the growth of single crystals. Nevertheless, due to the high significance of mass production of multicrystalline silicon (mc-Si) ingots for photovoltaics it will be briefly described here too.

Before the final configuration of a multigrain structure is completed it undergoes a dynamic process of ripening characterized by decay or dissolution of small nuclei and clusters in favor of larger nuclei or crystallites. This is due to the enhanced transport of atoms or molecules from the smaller to the larger objects initiated by the potential difference between them, as sketched in Fig. 5.14 (above right). Even if coalescence occurs the demolition of small grains is continued due to enhanced transport of atoms from fluid and even from smaller to larger grains leading finally to reduction of the total interface energy between the grains. This phenomenon was first described by *Ostwald* in 1897 and named after him as *Ostwald ripening*.

Wilhelm Ostwald
1853 - 1932

Chemical potential of large and small grain:

$$\Delta\mu_{large} = \Delta h_m \, \Delta T \, / T_m \, |_{r\to\infty}$$
$$\Delta\mu_{small} = \Delta h_m \, \Delta T \, / T_m + 2\gamma \, \Omega_V / r \, |_{r\to 0}$$
$$\Rightarrow \quad \Delta\mu_{large} < \Delta\mu_{small}$$

Snapshots of the time course in a wo-phase mixture

adapted from:
https://www.youtube.com/watch?v=IWJreldRjfs
see also: D. Fan et al., Acta Mater. 50 (2002) 1895

Final stage of an alum melt-solid interface started from randomly oriented nuclei

adapted from:
K. Ankit et al.,
Contrib. Mineral Petrol.
166 (2013) 1709

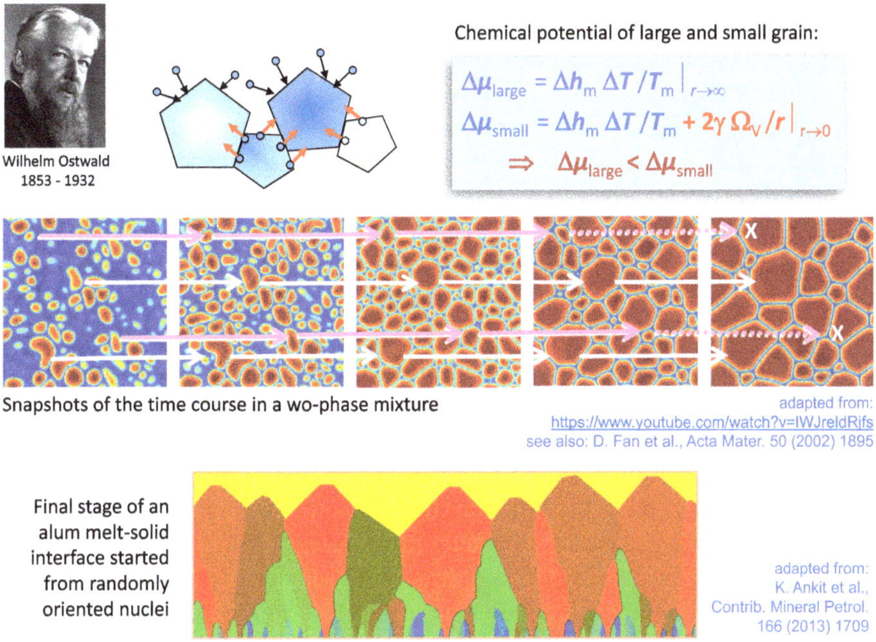

Fig. 5.14: Numeric 2D modeling of grain coarsening and Ostwald ripening (*the portrait of W. Ostwald is public domain; with permission of Elsevier (middle) and Springer-Nature (below)*).

The dominant source of such energy and mass transfer proves to be the relative cluster size. That means the Gibbs–Thomson effect (Section 4.2) becomes significant. As shown in eq. (4.10), the change of the chemical potential depends on the radius as

$$\Delta\mu = \frac{2\gamma\Omega_V}{r} \tag{5.42}$$

playing a decisional role even in small clusters whereupon a marked excess of energy is cumulated. Thus, the difference in chemical potential between clusters with small and large radius is

$$\mu_{r\to 0} = \mu_{r\to\infty} \exp\left(\frac{2\gamma\Omega_V}{rkT}\right) \tag{5.43}$$

Assuming nucleation of solid clusters from the vapor and replacing the chemical potential by the vapor pressure (see Spec box 3.3), eq. (5.43) can be expressed by the *Ostwald-Freundlich* (1909) equation

$$\frac{p_{r\to 0}}{p_{r\to\infty}} = \frac{p(r)}{p_{\text{eq}}} = \exp\left(\frac{2\gamma\Omega_V}{rkT}\right) = \exp\left(\frac{r_{\text{crit}}}{r}\right) \tag{5.44}$$

where $p(r)$ is the partial pressure at a curved interface of radius r (of the given cluster), p_{eq} the vapor pressure at flat interface (equilibrium pressure between the vapor and condensed phase), and $r_{\text{crit}} = 2\gamma\Omega_V/kT$. Thus, the shrinking of smaller clusters by reevaporation due to the higher partial pressure $p_{r\to 0} > p_{r\to\infty}$ is obvious. Then the releasing atoms can diffuse along the substrate surface to the larger clusters. When smaller clusters shrink to their critical nucleus size, they become thermodynamically unstable and spontaneously disintegrate.

In the middle of Fig. 5.14, the 2D-modeled time course of such process in a two-phase mixture is shown. After the nucleation within the fluid phase of overcritical crystallites with random orientation occurs their growth and coalescence takes place. When two particles with different orientations are in contact with each other, a grain boundary forms. Two particles will coalesce when they have the same orientation, which also happens in real systems. Further down the course, larger grains prevail (white arrows) and most small grains disappear (pink arrows).

An identical situation is obtained at growth from solutions where the chemical potentials are expressed by concentrations (or solubilities) and the higher supersaturation (solubility) of small clusters promotes the transfer of atoms or molecules from them to the larger crystallite by diffusion through the solution.

As was mentioned above, the ripening principle is continuing in a multigrain structure, i.e., when grains of different sizes are in contact with each other. At high enough temperature (let us say immediately after the nucleation process on the cooling bottom of a container for casting or directional solidification) the diffusive mass transfer from the smallest toward the largest grains occurs. This process is named *grain coarsening*. Again, the dominant source is the Gibbs–Thomson effect [see eq. (4.10)] promoting the shrinking of the smallest subgrains. Additionally, the *relative misorientation* proves to be another factor of grain mobility. The enhanced free energy per unit volume of atoms inside small grains with low radii of curvature produces a boundary migration (shrinking) toward its center of curvature. Finally, the boundary between the grains (grain boundary) is a high-entropy defect in the crystal structure and so it is associated with a marked excess of energy. The resulting thermodynamic driving force is acting toward reduction of the total area of boundary. This occurs, when the grain size increases, accompanied by a reduction in the actual number of grains per volume.

The local velocity of a grain boundary at any point is proportional to the local curvature of the grain boundary as

$$v_{GB} = M\,\gamma_{GB}\,\kappa \tag{5.45}$$

where M is the (kinetical) grain boundary mobility, γ_{GB} the grain boundary ("interface") energy, and κ the sum of the two surface curvatures. In comparison to the phase transformations the energy available to drive grain growth is very low and so it tends to occur at much slower rates and is easily slowed by the presence of second-phase particles or solute atoms in the structure.

Figure 5.14 below shows the longitudinal section of the 2D modeled final stage of grain growth completion of a free-grown alum crystal started from stochastic nuclei of various orientations. The growth competition results in the consumption of poorly oriented crystals (greenish and bluish in color) and in survival of favorably oriented ones (reddish). Further *geometrical selection* with propagating crystallization proves to cause the growth rate differences between differently oriented grains leading to mutual overgrowth. This effect takes place when at the fluid–solid interface the grains show faces of differing atomically flatness, i.e., of various free interface energies. According to this an atomically flat plane drops behind an atomically rough one (see lecture part II). Finally, grains oriented in the direction of highest thermal conductivity seem to be additionally favored at crystallization in a temperature gradient.

5.4 Nonequilibrium (kinetic) phase diagrams

As discussed in Section 5.1, a certain deviation from exact thermodynamic equilibrium is required to ensure crystallization by the driving force of crystallization or growth affinity. This holds true not only for the cases of nucleation of a (new) solid phase within a fluid one but also for the propagating fluid–solid interface of each crystallizing bulk or thin-film system. Therefore, the equilibrium phase diagrams treated in Chapter 3 do not exactly reflect the position of a growing interface. Of course, the usually acting supercoolings or supersaturations are relatively minimal, especially at the growth from melts. Also at the growth from solutions the crystallization velocities are so small that a near-equilibrium state can be approximated. However, the situation is changed at the growth under high supersaturations, like from the vapor phase, especially by the epitaxial methods MBE and MOCVD (see. Fig. 5.2) or in solidification processes of very high cooling and crystallization rates (e.g., casting). Strictly speaking, also at the single crystalline growth of materials showing atomically smooth interfaces, like most oxides and some semiconductors (see lecture part II), even at melt and melt–solution growth there is a characteristic kinetically determined retardation of the interface behind the equilibrium isotherm. Considering the deviation from equilibrium by undercooling or supersaturation *nonequilibrium (kinetic) phase diagrams* can be constructed.

Figure 5.15 shows three calculated T–x projections of nonequilibrium phase diagrams of a two-component system with the constituents A and B. The model developed by *Los and Matovic (2005)* describes the effective segregation taking place during the crystallization of solid–solutions from a binary liquid mixture, which incorporates the interfacial segregation and both mass and heat transport limitations in a coupled way. It is based on *nonequilibrium thermodynamics* (see Section 5.5), yielding a linear growth rate for each component and the theory of hydrodynamics with moving boundaries, assuming fixed boundary layers for both mass and heat transport (see lecture part III). It allows constructing the composition of the solid phase growing at nonequilibrium conditions for a given undercooling. Also the temperature and composition of the liquid phase at the interface can be read from these diagrams.

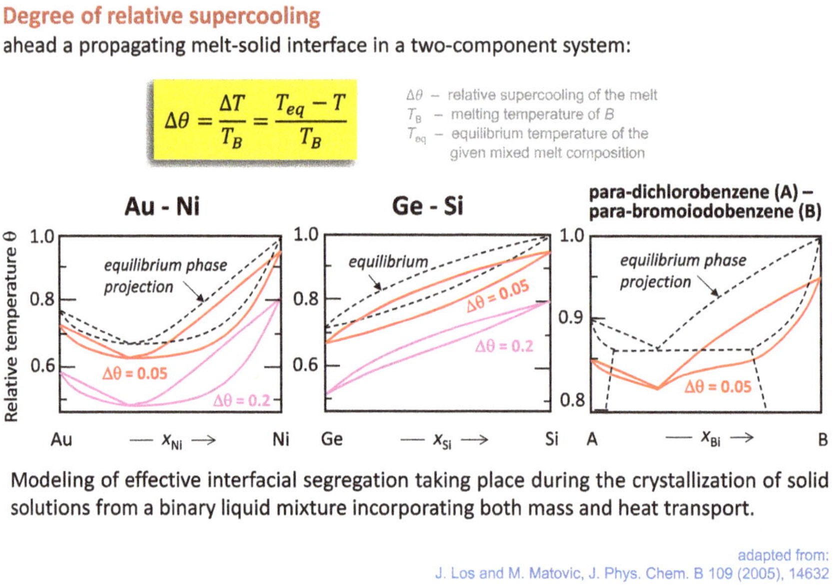

Fig. 5.15: Nonequilibrium (kinetic) phase projections (*with permission of Am. Chem. Soc.*).

The T-axis of the binary material examples shown is replaced by the dimensionless bulk liquid temperature $\theta = T/T_B$ with T_B the melting temperature of the component B with the highest melting temperature. Accordingly, the relative bulk undercooling $\Delta\theta$ is defined as

$$\Delta\theta = \frac{\Delta T}{T_B} = \frac{T_{eq} - T}{T_B} \tag{5.46}$$

where T_{eq} is the equilibrium temperature of the given melt–solid composition. As can be seen in comparison with the equilibrium phase diagrams the nonequilibrium projections are shifted downwards. The difference between the component concentrations in the liquid and solid phases (liquidus and solidus courses) are approaching in the nonequilibrium diagrams with increase of the undercooling (and also of the crystal growth rate). That means a kinetic distribution coefficient k_i can be deduced differing from the equilibrium coefficient k_0 introduced in Section 3.2.6.

Once again, in usual bulk single crystal processes the deviation from the phase equilibrium isotherm is minimal due to the usual low crystallization rates so that the equilibrium phase diagrams can be used as very good approximation. Nevertheless, the interest in adapting each growth step to the real nonequilibrium (kinetic) situation as precisely as possible is increasing, particularly for the study of segregation effect and the resulting distribution functions along the growing crystal. Progressive automation principles, such as *model-based control* of the crystal growth process, require such an approach.

5.5 Nonequilibrium thermodynamics: basic principles for crystal growth

As already emphasized in Section 1.2, each crystal growth system is characterized by import and export of heat fluxes. Further flow processes are taking place within the growing and cooling crystal caused by inhomogeneous component and thermal stress distributions evoking mass diffusion and elastic energy dissipations. Strictly speaking, each crystal growth arrangement is a "thermodynamically open system". Following the second law of thermodynamics nonequilibrium processes lead to a positive entropy production that never reaches zero as long as nonequilibrium is acting. Such situation can be described by principles of "*nonequilibrium (irreversible) thermodynamics*" whereupon the time dependence of the potential of Gibbs is irreversible

$$\frac{dG}{dt} = \frac{\partial H}{\partial t} - \frac{\partial S}{\partial t}T \quad \text{with} \quad \frac{\partial S}{\partial t} = P_S > 0 \tag{5.47}$$

where H and S are the system enthalpy and entropy, respectively, T is the absolute temperature, and P_s is the production of entropy not coming to a standstill. Figure 5.16 summarizes the phenomena of a crystal growth process, which strictly speaking make it necessary to treat it as a *thermodynamically open system.*

The total energetic *entropy production rate* combines the sum of the partial products of the driving thermodynamic forces F_i and flux density J_i for each dynamic event proceeding in both fluid phase and growing crystal

$$\frac{\partial S}{\partial t} = \sum F_i J_i \tag{5.48}$$

Therefore, in material systems the fluxes couple the entropy production with conductive and frictional processes responsible for energy dissipations. According to the *evolution criterion* of *Glansdorff and Prigogine* (1971) in such systems *self-organized states* designated as *dissipative structures* can be formed. This occurs when the export of entropy exceeds a critical value of the internal entropy production. It is clear that

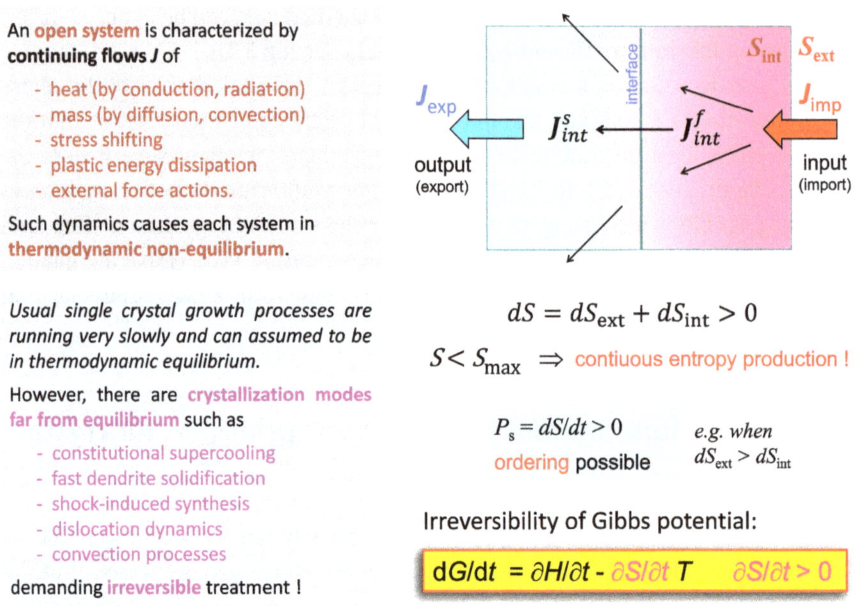
An open system is characterized by continuing flows J of
- heat (by conduction, radiation)
- mass (by diffusion, convection)
- stress shifting
- plastic energy dissipation
- external force actions.
Such dynamics causes each system in thermodynamic non-equilibrium.
Usual single crystal growth processes are running very slowly and can assumed to be in thermodynamic equilibrium.
However, there are crystallization modes far from equilibrium such as
- constitutional supercooling
- fast dendrite solidification
- shock-induced synthesis
- dislocation dynamics
- convection processes
demanding irreversible treatment !
interface
S_{int}
S_{ext}
J_{exp}
J_{imp}
J^s_{int}
J^f_{int}
output
(export)
input
(import)
$dS = dS_{ext} + dS_{int} > 0$
$S < S_{max}$ ⇒ contiuous entropy production !
$P_s = dS/dt > 0$
ordering possible
e.g. when
$dS_{ext} > dS_{int}$
Irreversibility of Gibbs potential:
$dG/dt = \partial H/\partial t - \partial S/\partial t\ T \quad \partial S/\partial t > 0$

Fig. 5.16: Principles of quasi-open crystal growth systems with flux export to be treated as thermodynamic nonequilibrium with irreversibility of Gibbs' potential.

such reduction of the inner entropy is associated with enhancement of ordering and, thus, of structuring possibility. Actually, dissipative structures are observed in crystal growth processes in the form of convection patterns in melt and solution, cellular arranged melt–solid interfaces, dendrites and dislocation cell patterns, for example. Mostly, they cause unwanted mesoscopic chemical and structural inhomogeneities. However, self-ordering may also prove useful especially in patterning of highly ordered nanocrystals by means of epitaxial processes. Further, steady-state periodic interface structures could help obtain periodic superstructures. In the following we will give an insight into the basics and possible applications of the irreversible thermodynamics in crystal growth. In this, the relative complex mathematical treatment will not be outlined in detail. For more information on this subject reference to the appropriate literature is made.

It must be underlined again that the phenomenological treatment of single crystal growth processes in the sense of equilibrium thermodynamics is quite acceptable due to the usually minimal deviation from phase equilibrium, more, when a good thermal isolation allows assuming them as quasi-closed systems. On the other hand there are numerous processes where the equilibrium is markedly exceeded, as in high phase potential differences (supercooling, supersaturation), very fast growth and cooling rates of the crystalline phase, well-developed convective conditions, thermoelastic stress dissipation in conjunction with high dislocation mobility, etc. (see Fig. 5.16).

The total inner system entropy $S = S_\Sigma$ consists of certain sub-entropic parts specified in Fig. 5.17. In the case of crystal growth these are the terms related to heat flow S_T, mass transfer S_i, friction S_η, and any internal (chemical) reactions S_Q, which in sum do not reach maximum entropy

$$S_\Sigma = S_T + S_i + S_\eta + S_Q + \cdots < S_{max} \tag{5.49}$$

Ilya Prigogine
1917 - 2003

According to the evolution criterion of *Glansdorff and Prigogine* in an irreversible open system any of the internal entropy parts can be removed.

As a result the sum of total entropy S_Σ decreases and a higher ordered state will appear referred to as <u>dissipative structures</u>

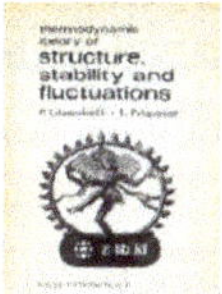

Paul Glansdorff
1904 - 1999

Total system entropy: $S_\Sigma = S_T + S_i + S_\eta + S_Q + \cdots < S_{max}$

Specific entropy production rate:

$$\frac{\partial s}{\partial t} = \sum F_i J_i = j_q \nabla \frac{1}{T} - \sum j_i \nabla \frac{\mu_i}{T} + \frac{1}{T}\sigma_\eta : \nabla v - \frac{1}{T}\sum \mu_i Q_i > 0$$

heat flux — mass flux — friction — chemical reactions

S_Σ - total system entropy, S_T, S_i, S_η, S_Q heat flow, mass transfer, friction, internal chemical reaction related sub-entropy part, respectively, t - time, F_i - thermodynamic driving force, J_i - flux density, j_q - heat flux density, T - absolute temperature, j_i - mass flux density, μ_i - chemical potential, σ_η - viscous stress tensor, v - velocity, Q_i - chemical production rate

adapted from: J. van der Eerden, Handbook of Crystal Growth Vol IA (Elsevier 1994);
see also: J. Kirkaldy, Canad. J. Physics 37 (1959)

Fig. 5.17: The evolution criterion of Glansdorff and Prigogine applied to thermodynamic nonequilibrium crystallization processes with continuous entropy production clarifies the appearance of dissipative structuring (*the portrait of I. Prigogine is public domain; with permission of Elsevier (formula)*).

According to eqs. (5.48) and (5.49), the specification of the time dependence of each sub-entropy part reveals the entropy balance equation (entropy production rate) in the intensive form

$$\frac{\partial s}{\partial t} = \sum F_i J_i = j_q \nabla \frac{1}{T} - \sum j_i \nabla \frac{\mu_i}{T} + \frac{1}{T}\sigma_\eta : \nabla v - \frac{1}{T}\sum \mu_i Q_i > 0 \tag{5.50}$$

with the heat, mass, friction (viscous dissipation) and chemical reaction-related entropy production parts (in the shown order of the right-hand side). Explicitly the terms mean: i) the heat conduction flux density $j_q = -\lambda \nabla T$ (λ – thermal conductivity) to the temperature gradient as driving force, ii) the sum of acting i particle flow densities (diffusivities) to the chemical potential gradients, iii) the (convection-related) viscous stress tensor σ_η (usually taken proportional to the viscous strain tensor ε as $\sigma_\eta = c{:}\ \varepsilon$ or in terms of tensorial product of the components $\sigma_{ij} = \Sigma c_{ijkl}{:}\ \varepsilon_{kl}$), which is the symmetrical part of the macroscopic velocity gradient ∇v, and iv) the internal chemical reaction productivity $\mu_i Q_i$ for small deviations from equilibrium (holding true in usual crystallization processes), equated with $K\sum k_r A$ (K – reaction constant, k_r – rate constant, A – reaction affinity = $-\sum v_i^r \mu_i / kT$ where v_i^r is the stoichiometric coefficient of species i in reaction r). Such chemical affinity can also refer to the tendency of an atom or molecule to combine by chemical reaction with atoms or molecules of unlike composition. Generally, when a chemical reaction is an irreversible process then it produces entropy. The minus sign stands for the decrease of thermodynamic potential as chemical reaction proceeds. Normally, in crystal growth processes any chemical reaction should be completed immediately at the interface and integrated in the latent heat of crystallization the removal of which is implied in the term of heat flux density. Nevertheless, during the cooling down process of multicomponent crystals any new reactions might become involved like complex formation and precipitation, for example.

How is a crystallization system able to self-organize dissipative structures and maintain its stable existence far from equilibrium? The main condition is its openness that enables continual flux of energy (see Fig. 5.16), into and out of a dissipative structure, which leads toward self-organization and ultimately the ability to function at a state of nonequilibrium. Figure 5.18 shows some images of self-organized dissipative structures being typical for crystal growth processes, such as the convective cell structuring within a melt heated from below (1), cellular interface formation when the criterion on morphological stability is neglected (2a,b), and dislocation cell patterning behind a growing interface (3a,b). We will come back to these phenomena and their origin parameters in parts III and IV when heat and mass transfer as well as defects are treated.

As per *Glansdorff and Prigogine* dissipative structures can exist in a stable stationary form because each deviation from the stability enlarges the entropy production according to $P_{\text{inst.}} > P_{\text{stab.}} \Rightarrow \partial S/\partial t \gg 0$ and, hence, enforces the repulsion of the structural perturbation bag to the steady state one. Actually, a growing melt–solid interface of cellular morphology is quite stable within a certain small parameter variations. This is also well-known from fast growing dendrites. Generally, in a multiflux and multiforces system the maintenance of the steady-state total entropy production requires the constancy of all acting partial entropy productions together as listed in eq. (5.49). Upon change of one of them the whole system reacts sensitively (note in the standard crystal growth practice this is rarely the case, but it could take place when, for instance, the friction related entropy part is changed by modifying an external energy input like magnetic Lorentz force). The sudden loss of one of the entropy parts would decrease the total entropy and, thus, an altered dissipative structure pattern of higher order (so called super-order) would be generated. Figure 5.19 demonstrates such an example. Two numerically calculated dendrite morphologies are formed in a supercooled Fe–B melt–solution under conditions far from equilibrium. For comparison the near-equilibrium shape of an iron crystal is sketched when growth at thermodynamic equilibrium would take place (Fig. 5.19a). Both dissipative structures were

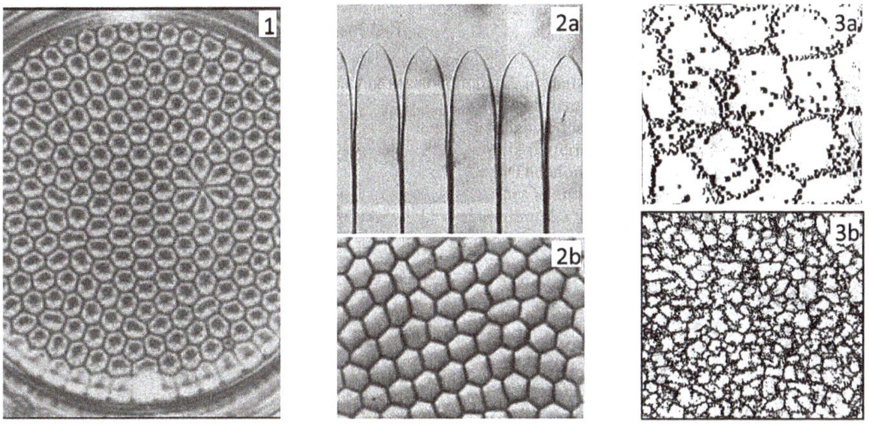

Hexagonal **convection pattern** (*Bernard* cells) in silicon oil heated from below (balance between thermal and viscous energy dissipation)

http://opencollaboration.wordpress (2010)
see also: E. Koschmieder, Beitr. Phys. Atmosph. 39 (1966) 1

Cellular interface formed at crystallization of metals; 2a - in situ top view, 2b - view on decanted interface (balance between heat and mass fluxes)

2a: B. Billia, HB CG IA (Elsevier1994)
2b: J. Rutter, Liqu. Met. Solidif. (1958)

Dislocation cell patterns in LiF (3a) and GaAs (3b) crystals (balance between elastic and plastic energy dissipation)

3a: G. Streb and B. Reppic, phys. stat. sol. (a) 16, 493 (1973);
3b: P. Rudolph, Progr. Cryst Growth Charact. Mat. 62 (2016) 89

Fig. 5.18: Characteristic self-organized dissipative structures with reference to crystal growth (*public domain of Wikipedia (1); with permission of Elsevier (2a, 3b), Taylor & Francisen (2b) and John Woley and Sons (3a)*.

grown from a crystallization center under differing transport conditions, in other words, by varying the number of entropy producing parts responsible for heat and mass transfer. In the case when the growth is controlled by both heat and mass transfer, almost smoothed (rounded) dendrite branches are formed (Fig. 5.19b). In comparison, when the growth is limited by diffusion of the dissolved component only and, thus, the total entropy production is reduced by the heat transfer related part, the dendrite becomes highly branched reflecting a super-ordered situation (Fig. 5.19c). Therefore, the removal of one or more partial entropy parts increases the degree of dissipative ordering.

Additionally, the ordered structure can adopt a steady-state oscillating mode. Such transitions have been widely studied in convective melt and solution volumes, especially in dielectric systems of low thermal conductivity. Varying states of order can even create *bifurcations*, which are unpredictable, irreversible, and extremely sensitive to change. That means the structure mode turns into branches (e.g., two oscillation frequencies appear when convection vortices are pulsating in diameter and rotation). After the bifurcation event, the system tends spontaneously to reorganize because each branch that was produced has less entropy, but in accordance with the second law of thermodynamics, in sum the same entropy as the ancestral situation.

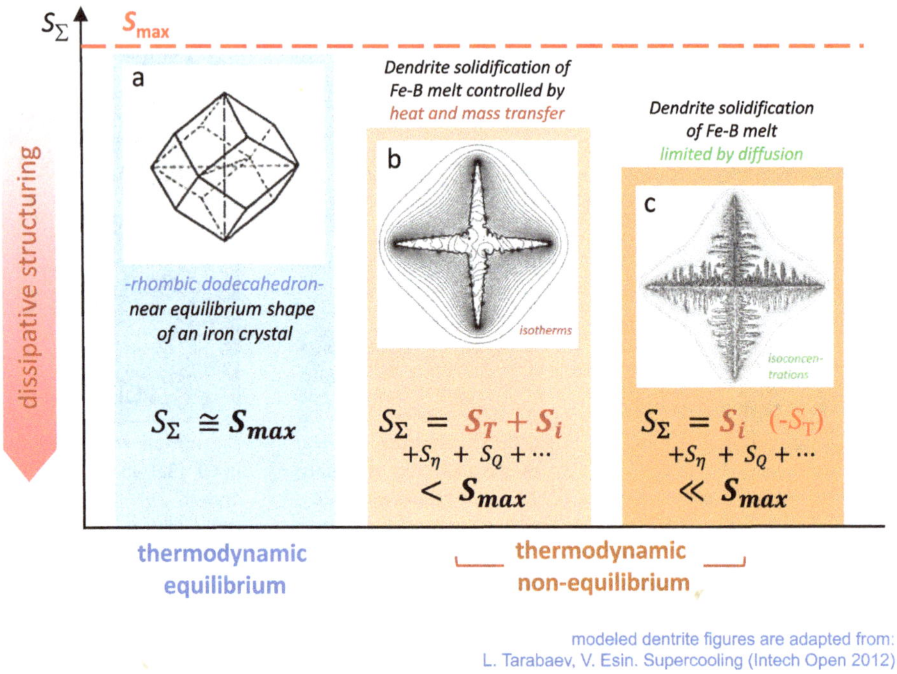

Fig. 5.19: Illustration of super-ordered dissipative patterning by removing of partial entropy parts from the total system entropy (*public domain of Wikipedia (a), open access of Intech Open (b, c)*).

As was already mentioned above, dissipative structures maintain their state by exchanging energy and matter constantly with the environment. This continuous interaction enables the system to establish an ordered structure with lower entropy than that of equilibrium state. A possible rhythmic structural situation can be described through the study of instabilities in nonequilibrium stationary states. One of the best-known physical ordering phenomena is the formation of convective *Benard cells* in a melt heated from below (approximately comparable with a Czochralski crucible) shown in Fig. 5.1. In 1900, *Benard* observed during heating from below of a thin fluid between two parallel horizontal plates that at a *critical temperature difference* the elevating effect of expansion predominates and the fluid starts to move in a structured way. It is divided into horizontally side-by-side arranged hexagonal convection cells in which the fluid rotates in a vertical plane. At the low plate the fluid is heated and rises but is cooled at the high plate leading to its density increase and movement downward. Such a structure needs continuous supply of energy and disappears as soon as the heating stops. The critical temperature difference ΔT_c can be determined from the critical dimensionless Rayleigh number acting in such a thin melt layer with quasi unlimited radial extensions

$$Ra_c = \frac{g\beta_T h^3}{\nu a}\Delta T_c = 1707 \tag{5.51}$$

where g is the gravitational acceleration, β_T the thermal expansion of the melt volume, h the distance between the plates, ν the kinematic viscosity, and a the thermal diffusivity $(=\lambda\ \rho^{-1}\ c_p{}^{-1})$. Therefore the critical temperature difference is

$$\Delta T_c = 1707\ \nu a\ \left(g\beta_T h^3\right)^{-1} \tag{5.52}$$

According to the hydrodynamic analysis the approximate velocity distribution in the *Benards cells* is

$$v(z) = \left[(Ra - Ra_c)/C\right]^{1/2} \cos\left(2\pi/\lambda_{\text{cell}}\right)^{-1} \tag{5.53}$$

where z is the vertical cell axis, C the constant, and λ_{cell} the repetition length of the cells aligned horizontally next to each other. The near hyperbolic cosine dependence between the velocity $V(z)$ and the Rayleigh number Ra reflect the bifurcation behavior at the critical value Ra_c at the curve minimum. That means, when the temperature difference is above a critical level, the resting fluid becomes unstable and it rotates in two structural states: one rotating toward the right and the other toward the left.

Another example of the application of the principle of nonequilibrium thermodynamics is the study of morphological stability of a cellular solidification front during unidirectional growth of a binary alloy by using the *Glansdorff–Prigogine formalism* generalized for surfaces (interfaces). Figure 5.20 shows three in situ images of a propagating steady-state melt–solid interface – one of flat morphology at near equilibrium conditions and the other two of cellular morphology at various deviations from thermodynamic equilibrium. The sketch presents a two-dimensional cell as a trigonometric function propagating along the z-axis with normal velocity v_n and having the cell periodicity (frequency) ω. Assuming that both phases S and L are infinite along the growth direction z and that the considered open thermodynamic system is limited between the boundaries z_S and z_L, then the wavy interface shape ϕ can be described by a cosine

$$\phi = \phi_0 \cos(\omega y) = \phi_0 \cos\left(\frac{2\pi}{\lambda_{\text{cell}}} y\right) \tag{5.54}$$

where ϕ_0 is the interfacial (cell) amplitude, y the coordinate along the interface, and $\lambda_{\text{cell}} = 2\pi/\omega$ the cell width. Now the question arises under which conditions such interface morphology remains stable or rather which growth parameters ensure a constant cell width during the growth process? In 1981 *Billia et al.* treated this problem mathematically and compared the results with experimental studies. The principle approach is the following.

First, viscous dissipation (convection within the melt) and chemical reactions are excluded. Thus, the entropy production according to eq. (5.50) is reduced to the heat- and mass-related parts

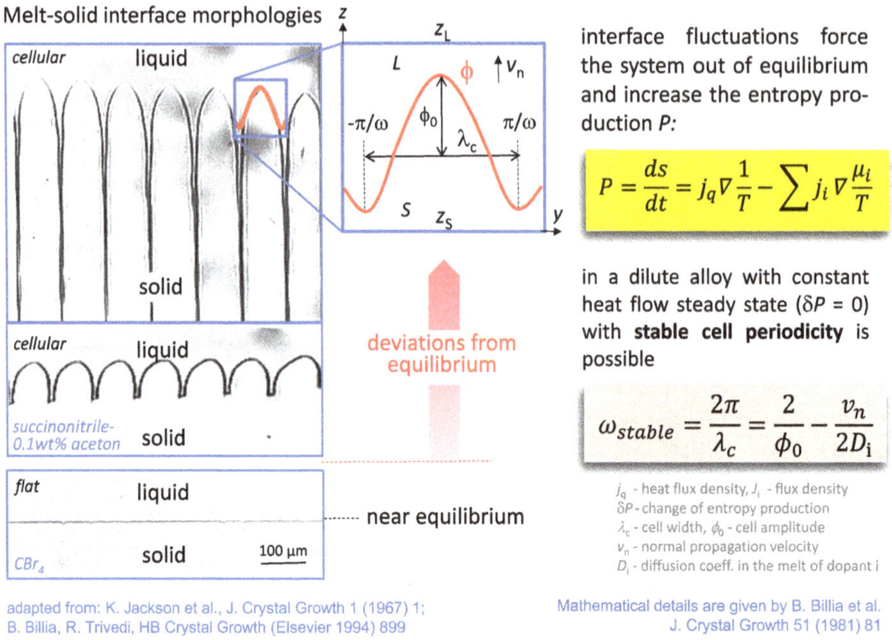

Melt-solid interface morphologies
z
cellular
liquid
solid
cellular
liquid
succinonitrile-0.1wt% aceton
solid
flat
liquid
CBr4
solid
100 µm
z_L
L
ϕ
v_n
-π/ω
ϕ_0
π/ω
λ_c
S
z_S
y
deviations from equilibrium
near equilibrium
interface fluctuations force the system out of equilibrium and increase the entropy production P:
$P = \frac{ds}{dt} = j_q \nabla \frac{1}{T} - \sum j_i \nabla \frac{\mu_i}{T}$
in a dilute alloy with constant heat flow steady state (δP = 0) with stable cell periodicity is possible
$\omega_{stable} = \frac{2\pi}{\lambda_c} = \frac{2}{\phi_0} - \frac{v_n}{2D_i}$
j_q - heat flux density, J_i - flux density
δP - change of entropy production
λ_c - cell width, ϕ_0 - cell amplitude
v_n - normal propagation velocity
D_i - diffusion coeff. in the melt of dopant i
adapted from: K. Jackson et al., J. Crystal Growth 1 (1967) 1; B. Billia, R. Trivedi, HB Crystal Growth (Elsevier 1994) 899
Mathematical details are given by B. Billia et al. J. Crystal Growth 51 (1981) 81

Fig. 5.20: Morphological stability of a cellular solidification front (*with permission of Elsevier (left images)*).

$$P = \frac{ds}{dt} = j_q \nabla \frac{1}{T} - \sum j_i \nabla \frac{\mu_i}{T} \tag{5.55}$$

Generally, for a steady-state (stable) interface shape ϕ, the related surface integral can be derived from the *variation principle* $\int_\phi P\, d\phi$ (see textbooks on stability analysis). Then the stability condition according the *Glansdorff–Prigogine* general evolution criterion is

$$\int_\phi \left[j_q \nabla \delta T_\phi^{-1} - \sum j_i \delta \left(\nabla \frac{\mu_i}{T} \right)_\phi \right] n_\phi d\phi \geq 0 \tag{5.56}$$

where δ stands for the variation (perturbation), n_ϕ is the normal to the interface in the growth direction, and T_ϕ is the temperatures of the interface being quasi the melting point of the alloy (note when this value is not in equilibrium with the melting point due to certain required undercooling as driving force of crystallization, as was treated in Section 5.1, the kinetic growth coefficient should be integrated; see lecture part II). Next the heat and mass flow densities are specified. The quantity of the heat flow through the growing interface is taken from the well-known equality $v_n \mathcal{L} \rho_S = \lambda_L \nabla_n T_L - \lambda_S \nabla_n T_S$ (see lecture part III) with $\mathcal{L}$ the latent heat of crystallization, ρ the density of solid, λ_L, λ_S, the thermal conductivities and $\nabla_n T_L$, $\nabla_n Ts$ the temperature gradients in the melt (L) and solid (S), respectively. The mass flow through the interface takes from the balance $D_i \nabla_n x_i = v_n x_i (k_0 - 1)$ with D_i the melt diffusion coefficient of the impurity i, x_i the mole fraction, and k_0 the equilibrium segregation coefficient (see Section 3.2.6). Therefore, eq. (5.56) becomes

$$\int_{\phi}\left[(\lambda_L \nabla_n T_L - \lambda_S \nabla_n T_S)\delta T_{\phi}^{-1} - \sum D_i \nabla_n x_i \delta\left(\frac{\mu_{1L}-\mu_{2L}}{T}\right)_{\phi}\right] n_{\phi} d\phi \geq 0 \tag{5.57}$$

where $\mu_{1\,L}$, $\mu_{2\,L}$ are the chemical potentials of the two components in the melt consisting of matrix 1 and impurity (dopant) 2.

Next is presumed that the heat flow balance is always constant without any perturbations so that the heat flow related term in eq. (5.57) can be neglected. This is quite legal when no oscillating convection is acting. In addition, the alloy is assumed to be diluted $x_{2L} \ll 1$. Finally, without detailed derivations, the integral within the boundaries $\pm\pi/\omega$ (see Fig. 5.20) is reduced on the stability of the concentration gradient in the melt along the cell contour

$$(\delta\omega/\omega)m \int_{-\pi/\omega}^{\pi/\omega} y(\partial x_{2L}/\partial y)dy \geq 0 \tag{5.58}$$

where $\delta\omega/\omega$ is the lateral perturbation of the cell width and $m = \partial T/\partial x_{21}$ is the slope of the liquidus in the T, x–phase projection. As the actual steady cellular state must be stable to any disturbance, it will correspond to the equality sign in eq. (5.58), which means that the entropy production is constant and, thus, $\delta P = 0$. After the functional expression of the mole fraction along the interface contour (here not shown) the stable cell periodicity $\omega = 2\pi/\lambda_{\text{cell}}$ is

$$\omega_{\text{stable}} = \frac{2}{\phi_0} - \frac{v_n}{2D_2} \tag{5.59}$$

with the cell amplitude $\phi_0 = D_2/v_n(k_{20}-1)A_E[1 + k_0 v_n/D_2(\bar{\omega} - v_n/D_2)]$, A_E is the coefficient characteristic of solute segregation in a cell, and $\bar{\omega} = v_n/2D_2\left[\left(v_n/2D_2{}^2 + n_{\phi}^2\omega^2\right)\right]$.

The comparison between theoretical and experimental cell sizes obtained by crystallization of Fe-8 wt% Ni alloys at various growth velocities showed quite a good conformity. For instance, at $v_n = 4.1$ cm/s and a temperature gradient of 31 K/cm experimental and theoretical cell widths yielded 500 µm and 410 µm, respectively.

There could be more examples like modeling of nonequilibrium phase diagrams already introduced in Section 5.4. These calculations are also based on principles of nonequilibrium thermodynamics whereupon the growth of binary mixed crystals from a liquid mother phase has been correlated to the *component fluxes* toward the solid phase (matrix A) and reverse from it (admixture B) depending on the degree of supercooling (deviation from thermodynamic equilibrium).

A further important phenomenon of nonequilibrium thermodynamics is the formation of dissipative cellular dislocation structures almost presented in mechanically or thermally stressed crystals (Fig. 5.18 3a and b). Even within cooling of as-grown crystals a strong dislocation dynamics takes place characterized by formation of globular cell arrangements. Such patterning requires in addition to the energetic nonequilibrium the balance between elastic and plastic energy dissipation.

Generally, there are numerous publications on nonequilibrium thermodynamics in material systems. However, according to the author's knowledge, until now a special handbook or even textbook on irreversible processes and dissipative structuring at crystal growth and epitaxy is still missing. Of course, first of all this has to do with the quasi-equilibrium conditions at growth of single crystals. Though, high-speed crystallization and epitaxial processes under high supersaturation are increasingly gaining significance. They correspond rather to nonequilibrium thermodynamics. In this context there is still a great deal of work ahead of us in order to develop a coherent theory on dissipative structuring, especially in thin-film growth processes.

6 Conclusions

We showed that thermodynamics is an important practical tool for crystal growth. It belongs not only to the basic theoretical knowledge of each crystal grower but also challenges its practical dexterity and clever modern mastering of process control and automation. The general principle of thermodynamics acing at each crystal growth process is the minimization of the free energy by returning from an excited situation to a lower energy state. Applied to the crystallization process, this means that the single crystalline state is a normal one because the *thermodynamic potential of Gibbs* of a solid phase becomes minimum if its building blocks (atoms and molecules) are perfectly packed in a three-dimensionally ordered crystal structure, i.e., the atomic bonds are saturated regularly. However, an ideally ordered crystalline state would imply a too limited entropy. Thus, the minimization of Gibbs free energy is also proportionally realized by an opposite directed force of increasing entropy causing *certain disorder.* As a result, both effects of opposite drives decrease the free energy as it is expressed mathematically by the equation of Gibbs $G = HT - S$. This *dialectics* will often meet at crystallization accompanied by defect generation.

Thermodynamics is a macroscopic science and, therefore, of phenomenological character only. It deals with average changes taking place among large numbers of atoms or molecules. It shows solely macroscopic start and end states, phase relations, *tendencies,* and *directions* but not the pathway in detail, as well as the microscopic steps of atomic size during the building of a crystalline structure. The transition from a fluid phase toward a crystalline state was discussed by the equilibrium treatment of *phase diagrams* of single- and multicomponent material systems. Special attention was given to the *region of homogeneity* of compounds and *stoichiometry* control, playing a decisive role in high-quality crystal growth. The *thermodynamic equilibrium segregation coefficients* for all phase transitions were derived.

The thermodynamics of surfaces, phase boundaries, and interfacial effects have been demonstrated to be of increasing importance even for epitaxial processes, multicrystalline solidification, and the growth of nanocrystals. It was shown how *faceting* influences the growth symmetry and meniscus stability when pulling from melt. In epitaxy, the *surface reconstruction* can contribute to *ordering effects* in mixed semiconductor thin films. Surface energy minimization can evoke *surface patterning* applicable in future nanostructuring.

Finally, we pointed out that the precondition for the crystallization of a stable solid phase within a metastable fluid phase is the *deviation from thermodynamic equilibrium.* We introduced the *driving force of crystallization,* which is required not only for nucleation processes but also acts at each propagating fluid–solid interface by a certain degree of *supercooling* or *supersaturation.*

At the end, nonequilibrium thermodynamics was introduced. Strictly speaking, each crystal growth arrangement is a "thermodynamically open system". Following

https://doi.org/10.1515/9783111711164-006

the second law of thermodynamics, nonequilibrium processes lead to a positive *entropy production* that never reaches zero as long as nonequilibrium is acting. Such a situation can be described by the principles of *nonequilibrium (irreversible) thermodynamics*, leading to the creation of dissipative structures. Although the standard crystal growth processes are running quasi-near to the thermodynamic equilibrium, some examples of typically stable nonequilibrium phenomena in bulk crystal growth, such as convection-driven patterning in melts or the formation of cellular interfaces in alloys, have been shown. We emphasized how important the consideration of nonequilibrium thermodynamics is during elastic energy dissipation in as-grown cooling crystals, leading also to dislocation cell patterning.

In summary, thermodynamics helps to understand "why" the crystalline phase is generated; however, the road of detailed atomistic steps leading to this, i.e., the "how" can be answered by the research field of *kinetics* only – our following lecture part.

Recommended literature

(With overview character and relation to crystallization; most of the author names are noted in italics in the book text)

Abraham, F.F., Homogeneous Nucleation Theory (Academic Press, New York 1974).

Allen, T.L., Shull, H., The Chemical Bond inMolecular Quantum Mechanics, J. Chem. Phys 35 (1961) 1644.

Becker, R. Döring, W., Kinetische Behandlung der Keimbildung in Übersättigten Dämpfen, Ann. Phys. 416 (1935) 719.

Bénard, H., Les tourbillons cellulaires dans une nappe liquide, Rev. Gén. Sci. Pure. Appl. 11 (1900) 1261 and 1309.

Billia, B., Ahdout H., Capella, L., Stable Cellular Growth of a Binary Alloy, J. Cryst. Growth 51(1981) 81; see also: Billia, B., Trivedi, B., Pattern formation in crystal growth, in: D.T.J. Hurle (ed.), Handbook of Crystal Growth, Vol. 1B (Elsevier, 1994) p. 899.

Boucher E.A., Jones, T.G., Capillary phenomena, J. Chem. Soc. 76 (1980) 1419.

Brebrick, R.F., Phase Equilibria, in: D.T.J. Hurle (ed.), Handbook of Crystal Growth, Vol.1A (Elsevier, Amsterdam 1993) p.43.

Brice, J.C., Crystal Growth Processes (Blackie, Halsted Press 1986).

Brice, J.C., Rudolph, P., Crystal Growth, in: Ullmann's Encyclopedia of Industrial Chemistry, Seventh Edition, 2007 Electronic Release (Wiley-VCH, Weinheim 2007).

Buhrig, E., Jurisch, M., Korb, J., Pätzold, O., Thermodynamic Modeling of Crystal Growth Processes, in: P. Capper, P. Rudolph (eds.), Crystal Growth Technology – Semiconductors and Dielectrics (Wiley-VCH, Winheim 2010) p. 3.

Campbell, F.C. (ed.), Phase Diagrams – Understanding the Basics (ASM International 2012).

Chiu, Cheng-Hsin, Gao, Huajian, Numerical simulation of diffusion controlled surface evolution, MRS Online Proceedings Library 317 (1993) 369.

Dalton, J. Essay IV. On the expansion of elastic fluids by heat, Memoirs of the Literary and Philosophical Society of Manchester 5 (1802) 595.

Chernov, A.A, Prozessy kristallisaziji, in: Sovremenaja kristallografija, t.3 (Izd. Moskva 1980).

Chernov, A.A., Modern Crystallography III, Crystal Growth (Springer, Berlin 1984).

Debye, P., Hückel, E., Zur Theorie der Elektrolyte, Phys. Z. 24 (1923) 185.

Dupré A., Théorie mécanique de la chaleur (Gauthier-Villard, Paris 1869).

Eerden, van der, J.P., Crystal Growth Mechanisms, ch. 2.4. Linear nonequilibrium thermodynamics, in: D.T.J. Hurle (ed.), Handbook of Crystal Growth, Vol. 1A (Elsevier, Amsterdam 1993) p.307.

Einstein, Th. L., Equilibrium shape of crystals, in: T. Nishinaga, P. Rudolph, T. Kuech (eds.), Handbook of Crystal Growth Vol. IA (Elsevier, Amsterdam 2015) p. 215.

Frank, F.C., van der Merwe, J.H., One-Dimensional Dislocations. II. Misfitting Monolayers and Oriented Overgrowth, in: Proc. Royal Society of London, Series A, Mathematical and Physical Sciences, vol. 198, (1949) 216.

Freundlich, H., Kapillarchemie (Akad. Verlagsges., Leipzig, 1909) pp. 144.

Gibbs, J.W., A Method of Geometrical Representation of the Thermodynamic Properties of Substances by Means of Surfaces, Trans. Conn. Acad., Vol. II (1873) 382.

Gibbs, J.W., On the Equilibrium of Heterogeneous Substances, Amer. Jour. Sci., ser. 3, vol. XVI (1878) 441.

Glansdorff, P., Prigogine, I., Non-equilibrium stability theory, Physica 46 (1970) 344.

Glicksman, M., Principles of Solidification (Springer, New York 2011).

Guisbiers, G., José-Yacaman, M. in: Encyclopedia of Interfacial Chemistry. Surface Science and Electrochemistry (Elsevier 2018) pp. 875.

https://doi.org/10.1515/9783111711164-007

Greenberg, J., Thermodynamic Basis of Crystal Growth – Phase P-T-X Equilibrium and Non-stoichiometry (Springer, Heidelberg 2002).
Guggenheim, E.A., Thermodynamics – An Advanced Treatment for Chemists and Physicists, 5th edition (North-Holland Publishing Comp., Amsterdam 1967).
Ham, F.S., Theory of diffusion-limited precipitation, J. Phys. Chem. Solids 6 (1958) 335.
Hayes, A., Chipman, J., Mechanism of Solidification and Segregation in a Low Carbon Rimming Steel Ingot Trans. AIME 135 (1939) 85.
Hein, K., Buhig E. (eds.), Kristallisation aus Schmelzen (Vlg. Grundstoffindustrie, Leipzig 1983).
Helmholtz, von, H., On the thermodynamics of chemical processes, in: Physical Memoirs, Selected and Translated from Foreign Sources (Taylor & Francis, Landon 1882).
Henry, W., Experiments on the quantity of gases absorbed by water, at different temperatures, and under different pressures,. Phil. Trans. R. Soc. Lond. 93 (1803) 29.
Herring, C., The use of classical macroscopic concepts in surface energy problems, in: Structure and Properties of Solid Surfaces, eds R. Gomer and C. S.Smith (University of Chicago Press, Chicago, 1953), pp. 5.
Hildebrand, J.H., A Quantitative Treatment of Deviations from Raoult's Law, Proc. Nat. Acad. Sci. 13 (1927) 267.
H. Hilton, Mathematical Crystallography and the Theory of Groups of Movements (Clarendon Press, Univ. of California, Oakland 1903).
Hollomon, J.H., Turnbull, D., Nucleation, Progress in Metal Physics 4 (1953) 333.
Ickert, L., Schneider, H.-G., Wachstum einkristalliner Schichten (VEB Dt. Vlg. Grundstoffindustrie, Leipzig 1983).
Jackson, K.A., Nucleation from the melt, Ind. Eng. Chem. 57 (1965) 29.
Jacobs, K., Fundamentals of Equilibrium Thermodynamics of Crystal Growth, in: H.J. Scheel, P. Capper (eds.), Crystal Growth Technology – From Fundamentals and Simulation to Large-Scale Production (Wiley-VCH, Weinheim 2008) p. 27.
Johansen, T.H., An improved analytical expression for the meniscus height in Czochralski growth, J. Cryst. Growth 141 (1994) 484.
Johnson, Ch. A., Generalization of the Gibbs-Thomson Equation, Surface Science 3 (1965) 429.
Kauzmann W., The nature of the glassy state and the behavior of liquids at low temperatures, Chem. Rev. 43 (1948) 219.
Kjelstrup, S., Røsjorde, A., Johannessen E., Non-Equilibrium Thermodynamics for Industry, in: Letcher, T. (ed.), Chemical Thermodynamics for Industry (Royal Soc. Chem. 2004).
Kirkaldy, J.S., Crystal Growth and the Thermodynamics of Irreversible Processes, Can. J. Phys. 37 (1059) 739.
Klimm, D. Phase Equilibria, in: T. Nishinaga, P. Rudolph, T. Kuech (Eds.), Handbook of Crystal Growth Vol. IA (Elsevier, Amsterdam 2015) p. 85.
Klimm, D., Thermodynamic and Kinetic Aspects of Crystal Growth, in: Handbook of Solid State Chemistry (Wiley-VCH, Weinheim 2017) p. 375.
Koga, Y., Solution Thermodynamics and Its Application to Aqueous Solutions (Elsevier, Amsterdam 2007).
Laplace P.S., Traite de Mechanique Celeste, Vol. 4 (Gauthier-Villars, Paris 1805)
Laue, von, M., Der Wulffsche Satz für die Gleichge-wichtsform von Kristallen Z. Kristallographie 105 (1943) 124.
Letcher, T., Teja A.S., Rousseau, R.W., Thermodynamics of crystallization, in: Letcher, T. (ed.), Chemical Thermodynamics for Industry (Royal Soc. Chem. 2004).
Liebmann, H., Der Curie-Wulff'sche Satz über Combinationsformen von Krystallen, Z. Kristallogaphie 53 (1914) 171.
Los, J.H., Matovic, M., Effective Kinetic Phase Diagrams, J. Phys. Chem. B 109 (2005) 14632.
Lupis, C.H.P., Chemical Thermodynamics of Materials (North-Holland Publ. Comp., New York 1983).

Machlin, S., An Introduction to Aspects of Thermodynamics and Kinetics Relevant to Materials Science: 3rd Edition (Elsevier, Amsterdam 2014).
Mahata, A., Mukhopadhyay, T., Zaeem, M.A., Liquid ordering induced heterogeneities in homogeneous nucleation during solidification of pure metals, J. Mater. Sci. Technol. 106 (2022) 77.
Markov I.V., Crystal Growth for Beginners (World Scientific, New Jersey 2003, 2008, 2020).
Mühlberg, M., Phase Diagrams for Crystal Growth, in: H.J. Scheel, P. Capper (eds.), Crystal Growth Technology – From Fundamentals and Simulation to Large-Scale Production (Wiley-VCH, Weinheim 2008) p. 3.
Mullin, J.B, Straughan B.W, Brickell W.S., Liquid encapsulation techniques: the use of an inert liquid in suppressing dissociation during the melt-growth of InAs and GaAs, J. Phys. Chem. Solids 26(1965) 782.
Nanev, Ch. N., Theory of Nucleation, in: T. Nishinaga, P. Rudolph, T. Kuech (Eds.), Handbook of Crystal Growth Vol. IA (Elsevier, Amsterdam 2015) p. 315.
Nishinaga, T., Thermodynamics – for Understanding Crystal Growth, Progress in Crystal Growth and Characterization of Materials 62 (2016) 43.
Ostwald, W., Studien über die Bildung und Umwandlung fester Körper 1. Abhandlung: Übersättigung und Überkaltung, Z. Phys. Chem. 22U (1897) 289.
Panofen, C., Herlach, D.M., Solidification of highly undercooled Si and Si–Ge melts, Mater. Sci. Eng. A 449–451 (2007) 699.
Pelton, A.D., Thermodynamics and Phase Diagrams of Materials, in: P. Haasen (ed.), Mat. Sci. and Technol. Vol.5 (VCH, Weinh.1991) p.1.
Pelton, A.D., Thompson, W.T., Phase Diagrams, Prog. Solid State Chem. 10 (1975) 119.
Pitzer, K.S., (Editor), Activity Coefficients in Electrolyte Solutions, Second Edition (CRC Press, Taylor and Francis Group, Boca Raton, FL 1991).
Raoult, F.-M., Recherches expérimentales sur les tensions de vapeur des dissolutions, J. Phys. Théor. Appl. 8 (1889) 5.
Redlich, O., Kister, A.T., Algebraic representation of thermodynamic properties and the classification of solutions, Ind. Eng. Chem. 40 (1948) 345.
Romanenko, V.N., Fizika tvordogo tela, T.2 (1960) 866; see also: K. Hein, E. Buhrig (eds.), Kristalliation aus Schmelzen (Vlg. Grundstaoffindustrie, Leipzig 1983) pp. 53.
Rosenberger, F., Fundamentals of Crystal Growth I (Springer, Berlin 1979).
Rudolph, P., Elements of thermodynamics for the understanding and design of crystal growth processes, in: R. Fornari, C. Paorici (eds.), Theoretical and Technological Aspects of Crystal Growth (Trans. Tech. Publ., Zurich 1998) p.1.
Rudolph, P., Thermodynamic fundamentals of phase transition applied to crystal growth processes, in: H.J. Scheel, T. Fukuda (eds.), Crystal Growth Technology (John Wiley & Sons Ltd., West Sussex 2003) p. 15.
Rudolph, P., Stoichiometry related growth phenomena and methods of controlling, in: K. Byrappa, H. Klapper, T. Ohachi, R. Fornari (eds.), Crystal Growth of Technologically Important Electronic Materials (Allied Publishers Pvt., New Delhi 2003) p. 407.
Sadeghi M., Rasmuson Å.C., On the estimation of crystallization driving forces, Cryst. Eng. Comm. 21 (2019) 5164.
Satunkin, G.A., Determination of growth angles, wetting angles, interfacial tensions and capillary constant values of melts, J. Cryst. Growth 255 (2003) 170.
Schmelzer J.W.P. (ed.), Crystallization Thermodynamics, Special Issue of Entropy – Open Access Journal by MDPI (MDPI, Basel 2019).
Schultz, P.A., First-principles calculations of metal surfaces. I. Slab-consistent bulk reference for convergent surface properties, Phys. Rev. B 103 (2021) 195426.
Sekerka, R.F., Thermal Physics: Thermodynamics and Statistical Mechanics for Scientists and Engineers (Elsevier, Amsterdam 2015).

Srolovitz D.J., On the stability of surfaces of stressed solids, Acta Metal. 37 (1989) 621.
Steinhardt, P.J., Nelson D.R., Ronchetti M., Bond-orientational order in liquids and glasses, Phys. Rev. B28 (1983) 784.
Stranski, I.N.; Krastanow, L., Zur Theorie der orientierten Ausscheidung von Ionenkristallen aufeinander, Abhandlungen der Mathematisch-Naturwissenschaftlichen Klasse IIb, Akad. der Wiss. Wien 146 (1938) 797.
Stringfellow, G.B., Thermodynamics of Modern Epitaxial Growth Processes, in: G. Müller, J.-J. Métois, P. Rudolph (eds.), Crystal Growth – From Fundamentals to Technology (Elsevier, Amsterdam 2004) p. 1.
Stringfellow, G.B., Thermodynamic aspects of organometallic vapor phase epitaxy, J. Cryst. Growth 62 (1983) 225.
Stroiteljew, S.A., Kristallokhimitcheskiij aspekt tekhnologii poluprovodnikov (Izd. Nauka, Sib. Otdelenije. Novosibirsk 1976)
Thomson, J.J., Applications of dynamics to physics and chemistry (Macmillan and Co., London 1888).
Turnbull, D., Formation of crystal nuclei in liquid metals, J. Appl. Phys. 21(1950) 1022.
Vegard, L., Die Konstitution der Mischkristalle und die Raumfüllung der Atome, Z. Phys. 5 (1921) 17.
Vekilov, P.G., The two-step mechanism of nucleation of crystals in solution, Nanoscale 2 (2010) 2346.
Vigdorovich, V.N., Ochistka metallov i poluprovodnikov kristallizazieij (Metallurgija, Moskva 1969).
Volmer, M., Zum Problem des Kristallwachstums, Zs. Phys. Chem. 102 (1922) 267.
Volmer, M., Weber, A., Keimbildung in übersättigten Gebilden, Zs. Phys. Chem. 119 (1926) 277.
Voronkov, V.V., O termitcheskom ravnovesii na linii razdela trekh faz, Sov. Solid. St. Phys. 5 (1963) 571c.
Weiser, K., Theoretical calculation of distribution coefficients of impurities in germanium and silicon, heats of solid solution, J. Phys. Chem. Solids 7 (1958) 118.
Wilke, K.Th., Bohm, J., Kristallzüchtung, Ch. 1.2.1. Zustandsdiagramme u. treibende Kräfte (Vlg. d. Wiss. Berlin 1988 and H. Deutsch and Thun, Frankfurt a.M. 1988).
Wilkes, J.G., The precipitation of oxygen in silicon, J. Cryst. Growth 65 (1983) 214.
Wilson, P. (ed.), Supercooling (InTechOpen 2012).
Wood E.A., Vocabulary of surface crystallography. J. Appl. Phys. 35(1964) 1306.
Wulff, G.V., Zur Frage der Geschwindigkeit des Wachsthums und der Auflösung der Krystallflächen, Zs. Kristallogr. Mineral. 34 (1901) 449.
De Yoreo, J.J., Vekilov, P.G., Principles of Crystal Nucleation and Growth, Rev. Mineral. Geochem. 54 (2003) 57.
Zeldovich, Y.B., On the theory of new phase formation: cavitation, Acta Physicochim. URSS 18 (1943) 1.
Zhang, Ke-Qin, Liu, Xiang Y., In situ observation of colloidal monolayer nucleation driven by an alternating electric field, Nature 429 (2004) 739.
Zunger, A., Spontaneous atomic ordering in semiconductor alloys: causes, carriers, and consequences, MRS Bull. 22 (1997) 20.

Abbreviations

AFM	Atomic force microscopy
BR	Bridgman (method)
CNA	Common neighbor analysis
CNT	Classical nucleation theory
CMP	Congruent melting point
CRSS	Critical resolved shear stress
CZ	Czochralski (method)
DFT	Density functional theory
DSC	Differential scanning calorimetry
DTA	Differential thermal analysis
EFG	Edge-defined film-fed growth (method)
FZ	Floating zone (method)
HBR	Horizontal Bridgman (method)
HVPE	Hydride vapor-phase epitaxy (method)
IF	Interface
LEC	Liquid encapsulation Czochralski (method)
LEED	Low-energy electron diffraction
LPE	Liquid-phase epitaxy (method)
MBE	Molecular beam epitaxy (method)
MOCVD	Metal-organic chemical vapor deposition (method)
OP	Order parameter
PD	Phase diagram
RHEED	Reflection high-energy electron diffraction
SPA	Surface photoabsorption
STM	Scanning tunneling microscopy
THM	Traveling heater method (≡ traveling solvent zone) (method)
TSSG	Top seeded solution growth (method)
VB	Vertical Bridgman (method)
VCz	Vapor pressure-controlled Czochralski (method)
VGF	Vertical gradient freeze (method)
VPE	Vapor-phase epitaxy (method)
W	Wurtzite (structure)
ZB	Zinc blende (structure)
µ-PD	Micropulling down (method)
2D, 3D	Two-dimensional and three-dimensional

https://doi.org/10.1515/9783111711164-008

Spec boxes

Spec box 2.1 The heat quantity —— **10**
Spec box 2.2 The enthalpy of solution —— **13**
Spec box 3.1 The substance indication —— **20**
Spec box 3.2 The mole fraction in systems with compounds —— **22**
Spec box 3.3 The chemical potential of ideal and real solution —— **24**
Spec box 3.4 The entropy of mixing —— **26**
Spec box 3.5 The tangent construction —— **30**
Spec box 3.6 Predictions of the interaction parameter —— **38**
Spec box 3.7 Growth velocity at stoichiometry-controlled vertical Bridgman technique —— **57**
Spec box 3.8 Extra source parametrization at stoichiometry-controlled melt growth —— **60**
Spec box 3.9 Theoretical approaches to the equilibrium segregation coefficient —— **77**
Spec box 4.1 Derivation of classic Gibbs-Thomson equation —— **95**
Spec box 4.2 The constancy of h_j/γ_j ratio —— **100**
Spec box 5.1 Driving forces of crystallization —— **127**
Spec box 5.2 The critical solid nucleus of polyhedral shape —— **139**

https://doi.org/10.1515/9783111711164-009

Index

activity coefficient 22, 23, 25, 27, 33–34, 36, 38–39, 42, 75, 77, 90, 128
atomistic ordering 117
axial distribution 87
axial dopant distribution 86, 87

Benard cells 168, 169
berthollide 66–67
binary material compounds 21
Bridgman method 51, 53, 55–57, 60, 62–63, 111

capillary constant 107, 108
Clausius-Clapeyron equation 59, 61
component redistribution 74
compound 3, 13, 17, 20–22, 24, 35, 39, 42–57, 59, 63–72, 74, 82–87, 94, 111, 114, 115, 120, 154, 156, 157, 173
compound existence region 64, 85
configurational entropy 25, 27, 32, 78
congruent evaporation 50
congruent melting point (CMP) 44–46, 50–54, 68, 82, 83
constitutional supercooling 42, 55, 67
crystal habitus 111
Czochralski method 10, 11, 49, 63–65, 106, 107, 109, 111, 133, 156, 168

daltonide 66
density functional theory 94
differential scanning calorimetry (DSC) 9–10
diffuse interface 144
diffusion boundary layer 42, 55–59, 62, 64, 84
diluted solution 81, 82
disk-shaped nucleus 149
dissipative structures 6, 163–168, 172, 174
distribution coefficient 73–75, 82, 85, 87, 89, 163
driving force of crystallization XIV, 3, 125–130, 140, 143, 145, 161, 170, 173

EFG method 106
elastic strain energy 129
enthalpy of mixing 24, 27, 32–35, 39
entropy of mixing 24–27, 32, 33
epitaxial layers XI, XII, XIII, 40, 72, 106
equilibrium distribution coefficient 41, 60, 73, 85, 87, 152
equilibrium shape of crystals 91, 97–106, 166

Euler's theorem 21
eutectic 20, 35, 45, 65, 67, 70–72, 81, 82
evolution criterion of Glansdorff and Prigogine 6, 163, 165

facets XII, 94, 106, 116, 123, 152, 155
facetting 108–112, 173
first law of thermodynamics 7
first-order phase transition 8, 9, 11–12, 15, 67, 130
floating zone method 106, 108
Frank-van der Merwe mechanism 151
fugacity 22, 33, 48

Gibbs free energy 7–8, 16, 20, 21, 26, 31, 36, 44, 65, 67, 92, 97–98, 104, 125, 127, 136, 146, 173
Gibbs' phase rule 16, 17, 65
Gibbs potential 20, 164
Gibbs surface potential 91
Gibbs-Thomson effect 153, 159, 160
Gibbs-Thomson equation 95–98, 136
Glansdorff-Prigogine formalism 169, 170
glass formation 15, 142, 143
grain boundary 160–161
grain coarsening 158–161
growth affinity 125, 165
growth angle 106–108, 110
G-T-p surfaces 16, 17

heat of crystallization 9, 10, 13, 166, 170
heterogeneous nucleation 132, 133, 145–148, 151, 154, 158
homogeneity range 54
homogeneity region 44–46, 156
homogeneous nucleation 128, 130, 133–145, 147–148, 158

ideally mixed phases 28
incongruent evaporation 48–53
incongruent melting 65–67
incongruent melting point 46, 66
industrial crystallization 132
intensive property 20, 100
interaction parameter 34, 36–39, 42, 71, 75, 77–79, 128
irreversible thermodynamics 5, 163, 164, 174
Ising model 104
isomorphous system 70

https://doi.org/10.1515/9783111711164-010

kinetic Wulff plots 106
Kyropoulus method 55, 89

Laplace equation 106
latent heat of crystallization 9, 10, 166
liquid encapsulant 49, 53, 63–65
LPE (liquid-phase epitaxy) 71, 72, 85, 102, 121, 123, 129, 130

MBE (molecular beam epitaxy) 113, 115, 116, 123, 125, 130, 161
melt-glass transition 143
melt-solutions 11, 13–14, 22, 67, 70, 72, 84, 86–88, 129, 130, 150, 161, 166
Miller indices 98, 108, 139
mixed crystal 21–23, 26, 27, 31, 33, 35, 38–42, 70, 72, 81, 85, 87, 89, 120, 122, 171
mixed system 20–22, 26–40, 42, 49, 61, 63, 74, 77, 118, 120
MOCVD (metal-organic chemical vapor phase deposition) 89, 119, 123, 130, 161
model of quasi-chemical equilibrium 34, 38, 78
molar free energy 23, 30
molar Gibbs free potential 20, 125
multicrystalline structure 132
multiple material compounds 21

nanocrystal XI, XII, XIII, 97, 98, 102, 103, 106, 123, 124, 133, 164, 173
nanocrystal growth 97
nanowire 123
non-equilibrium phase diagram XV, 161–163
non-equilibrium thermodynamics 4–6, 110, 152–154, 163–174
nucleation XVI, 3, 46, 63, 69, 74, 103, 106, 109–112, 117, 125, 128, 130–161, 173
nucleation rate 137, 138, 143, 153–154

ordering effect 113, 117–120, 144, 173
order parameter 144
Ostwald ripening 132, 158–161
Ostwald's step rule 134

partial vapor pressure 33, 48, 61, 88, 90
peritectic line 65, 66
polymorphism 68
precipitate 46, 51, 65, 156–158
precipitation 156–158, 166
purification effect 85, 87
p-V projections 17, 19

quantity of heat 10, 11
quantum dots 121

random mixing of components 32
Raoult's Law 23, 24, 33, 48, 60, 88
Raoult's relations 48, 57
rate of nucleation 152
real mixed system 31–40, 49
reconstruction 12, 91, 104, 112–117, 119, 173
Redlich-Kister equation 36
region of existence 44
regression functions 80, 81
regular solution 31–33, 75
ridge-like protrusions 108–112

seed crystal 22, 68, 69, 133
segregation 29, 31, 41, 42, 45, 52, 53, 55–57, 73–90, 119, 162, 163, 171
self-assembling 113, 120–124
self-organized states 163, 167
singular faces 101, 102, 109
solid-liquid interface 42, 64, 106–108, 142, 154
solid-solid phase transition 10–12, 67–69
stoichiometric composition 43–46, 50–54, 56, 57, 59–61, 65–67, 74
stoichiometry 17, 20, 42–48, 50–67, 74, 82–83, 86, 89, 105, 156, 166, 173
Stranski-Krastanov mode 121, 123, 151
sub-regular model 36
substrate XI, XII, 40, 66, 70, 71, 106, 112, 115–124, 128, 130, 132–133, 145–147, 149, 150–152, 154, 160
supercooling 3, 9, 15, 42, 55, 67, 94, 105, 125, 128, 130–131, 133, 134, 138–144, 151, 161, 164, 166, 173
surface-active substances 93
surface energy 91–96, 98–124, 131, 143, 149–151, 173
surface of a crystal 91, 113
surface slab 94, 95
surface tension 91–94, 106, 148, 151
surfactants 104, 105, 120
symmetric solution model 34–36, 128
system of ideal mixed solutions 75

tangent construction 29–30, 43
temperature-time-transformation (TTT) 158
ternary systems 70–73
thermal differential analysis 9–11, 92, 95, 96, 107
thermodynamically open system 5, 163, 173
THM (traveling heater method) 67, 70, 72, 85–88, 110, 133, 168
T-p projection 16, 17, 19, 54, 115
triple phase line 17, 19, 108, 109, 134
Turnbull's rule 94
two-phase state 30, 72
T-x projections 29, 31, 36, 40–42, 44, 50, 53, 54, 65, 66, 74, 76, 79–82, 161

undercooling 96, 97, 109, 110, 125, 128, 136, 141–143, 148, 155, 161, 162, 170

van Laar equation 41
Vegard`s rule 42
vertical gradient freeze (VGF) 53, 55, 56, 59, 62–65, 69, 89, 111, 133, 154, 155
vicinal faces 101, 102

waved surface profile 122
wetting angle 107, 145–148, 155
wetting function 147, 148
Wulff's plot 100
Wulff's theorem 99

Young's Equation 94, 145

Zeldovich factor 138, 153

www.ingramcontent.com/pod-product-compliance
Lightning Source LLC
Chambersburg PA
CBHW081524220326
41598CB00036B/6320
9783111711058